Progress in Mathematical Physics
Volume 61

Editors-in-Chief

Anne Boutet de Monvel, *Université Paris VII Denis Diderot, France*
Gerald Kaiser, *Center for Signals and Waves, Austin, TX, USA*

Editorial Board

C. Berenstein, *University of Maryland, College Park, USA*
Sir M. Berry, *University of Bristol, UK*
P. Blanchard, *University of Bielefeld, Germany*
M. Eastwood, *University of Adelaide, Australia*
A.S. Fokas, *University of Cambridge, UK*
D. Sternheimer, *Université de Bourgogne, Dijon, France*
C. Tracy, *University of California, Davis, USA*

Glasses and Grains

Poincaré Seminar 2009

Bertrand Duplantier
Thomas C. Halsey
Vincent Rivasseau
Editors

 Birkhäuser

Editors

Bertrand Duplantier
Service de Physique Théorique
Orme des Merisiers
CEA – Saclay
91191 Gif-sur-Yvette Cedex
France
bertrand.duplantier@cea.fr

Thomas C. Halsey
ExxonMobil Upstream Research Company
3120 Buffalo Speedway
Houston, TX 77098
USA
thomas.c.halsey@exxonmobil.com

Vincent Rivasseau
Laboratoire de Physique Théorique
Université Paris-Sud
Campus d'Orsay
91405 Orsay Cedex
France
Vincent.Rivasseau@th.u-psud.fr

ISBN 978-3-0348-0083-9   e-ISBN 978-3-0348-0084-6
DOI 10.1007/978-3-0348-0084-6

Library of Congress Control Number: 2011928087

2000 Mathematics Subject Classification: 76-xx, 80-xx, 82-xx, 35-xx, 37-xx, 60-xx, 65-xx

Cover design: deblik, Berlin

Printed on acid-free paper

Springer Basel AG is part of Springer Science+Business Media

www.birkhauser-science.com

Contents

Yoël Forterre and Olivier Pouliquen
Granular Flows

Thomas C. Halsey
Theoretical Considerations for Granular Flow

Olivier Dauchot
Grains, Glasses and Jamming

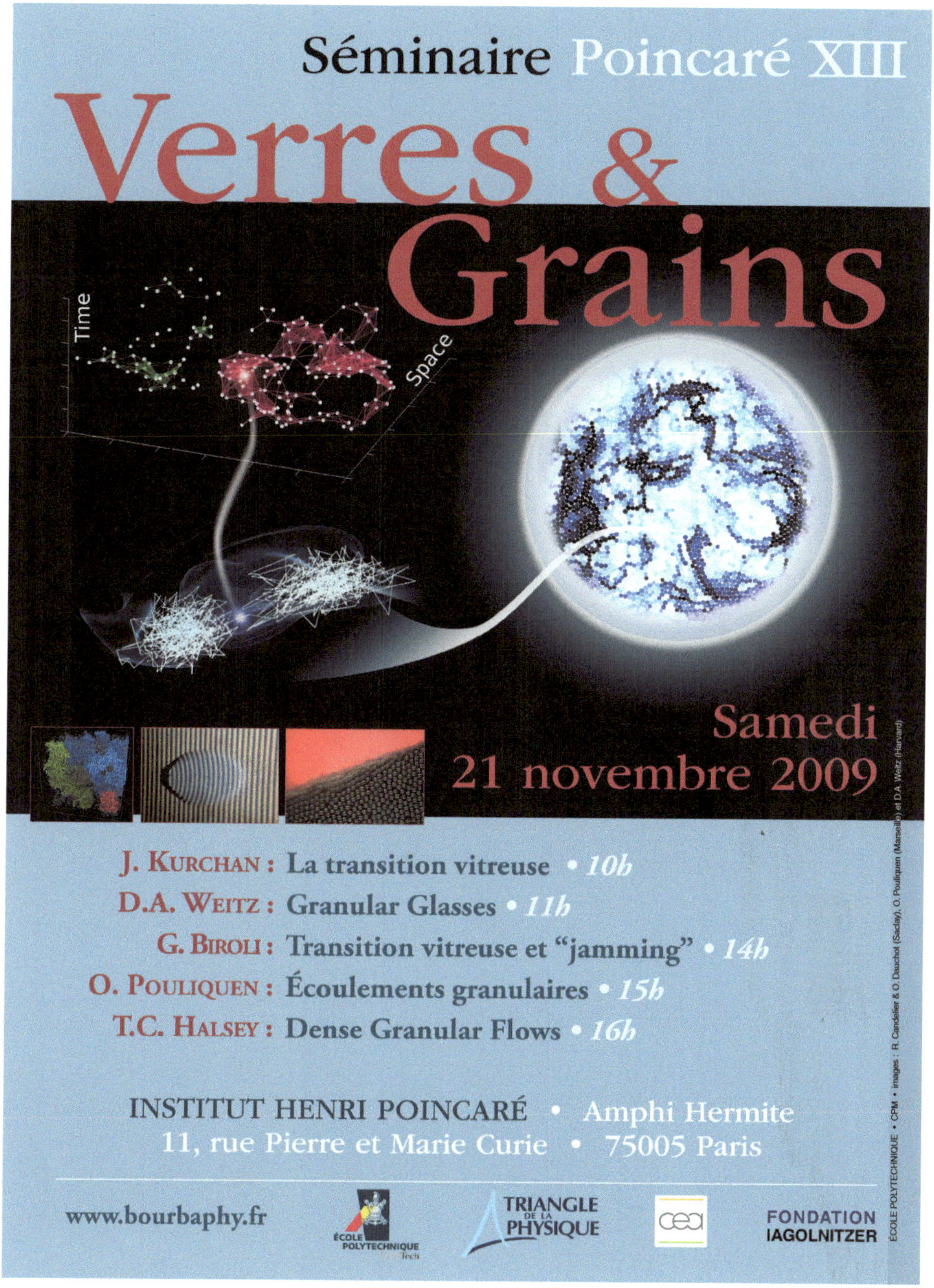

Séminaire Poincaré XIII

Verres & Grains

Time

Space

Samedi
21 novembre 2009

J. KURCHAN : La transition vitreuse • *10h*

D.A. WEITZ : Granular Glasses • *11h*

G. BIROLI : Transition vitreuse et "jamming" • *14h*

O. POULIQUEN : Écoulements granulaires • *15h*

T.C. HALSEY : Dense Granular Flows • *16h*

INSTITUT HENRI POINCARÉ • Amphi Hermite
11, rue Pierre et Marie Curie • 75005 Paris

www.bourbaphy.fr

ÉCOLE
POLYTECHNIQUE
Tech

TRIANGLE
DE LA
PHYSIQUE

cea

FONDATION
IAGOLNITZER

ÉCOLE POLYTECHNIQUE • CPM • images : R. Candelier & O. Dauchot (Saclay), O. Pouliquen (Marseille) et D.A. Weitz (Harvard)

Foreword

This book is the tenth in a series of Proceedings for the *Séminaire Poincaré*, which is directed towards a large audience of physicists, mathematicians, and biologists.

The goal of this Seminar is to provide up-to-date information about general topics of great interest in physics. Both the theoretical and experimental aspects of the topic are covered, generally with some historical background. Inspired by the Nicolas Bourbaki seminar in mathematics, hence nicknamed "Bourbaphy", the Poincaré Seminar is held twice a year at the Institut Henri Poincaré in Paris, with written contributions prepared in advance. Particular care is devoted to the pedagogical nature of the presentations, so as to fulfill the goal of being useful to a large audience of scientists.

This new volume of the Poincare Seminar Series, *"Glasses and Grains"* ("Verres et Grains"), corresponds to the thirteenth such seminar, held on November 21, 2009. It describes recent developments in the statistical physics of two related but still poorly understood topics – glasses, especially the glass transition, and the statics and dynamics of granular systems. This field has emerged as one of the most challenging frontiers of statistical physics in the last two decades, and is notable for its very active interchange between experiment, theory, and numerical studies.

The first survey, by JORGE KURCHAN, simply titled *"Glasses"*, summarizes the elements of the low temperature and transition behavior of glasses. He emphasizes the collective nature of glassy "order", as implied by the appearance of non-Arrhenius relaxations; he also discusses aging, the significance of density-functional approaches, and finally gives a brief introduction to the Random First Order mean-field theory, which has refocused theoretical approaches to the understanding of the glass transition.

The second article, *"Colloidal Glasses"*, by DAVE WEITZ, presents an experimental perspective on the glass problem, by focusing on colloids, for which glass-forming occurs even with relatively simple inter-particle interactions. Repulsive colloidal glasses seem to be highly analogous to granular materials, exhibiting a "jamming" transition in which a correlation length *appears* to grow without limit. The behavior of attractive colloidal glasses is more complex, showing a spinodal decomposition followed by a kinetic gelation as the attraction is increased.

In the third contribution, *"Glass and Jamming Transitions"*, GIULIO BIROLI returns to the theory of the glass transition, discussing in greater detail the Random First Order Transition theory originally proposed by Kirkpatrick, Thirumalai, and Wolynes, and introduced in the first contribution to this volume. With the help of a set of witty Galilean dialogues, Biroli summarizes both the seminal nature of this breakthrough as well as its limitations as a full theory of the transition and low-temperature properties of glasses.

The fascinating phenomenology of granular flows is then discussed by YOEL FORTERRE and OLIVIER POULIQUEN in *"Granular Flows"*. After a brief review of the distinctive features of granular flows vis-a-vis liquid flows, they review recent work emphasizing the role of the "Inertial Number", measuring the ratio of inertial forces to internal stresses, in controlling flow rheology. They then summarize a recently proposed visco-plastic model, based on this insight, which organizes a broad range of results for dense granular flows in a simple, intuitive way.

Finally, one of the editors of this volume, THOMAS HALSEY, examines the underlying microscopic mechanisms that are expressed through this Inertial Number dependence in *"Theoretical Considerations for Granular Flow"*. He presents an exact solution for the flow of a highly symmetric "honeycomb" granular packing, extends these results to random packings, and proposes an interpretation, based on an underlying length scale in the flow, for the Inertial Number dependence summarized in the previous contribution.

In addition to the above contributions, which were presented at the original seminar in November 2009, the editors have completed this volume by obtaining a further review entitled *"Grains, Glasses, and Jamming"*, by OLIVIER DAUCHOT. Dauchot focuses on the analogy between glassy relaxation and the relaxation of granular packings under mechanical excitation, exploring similarities and differences in the spatio-temporal organization of relaxation modes in these two systems.

We hope that the continued publication of this series of Proceedings will serve the scientific community, at both the professional and graduate levels. We thank the COMMISSARIAT À L'ÉNERGIE ATOMIQUE (Division des Sciences de la Matière), the DANIEL IAGOLNITZER FOUNDATION, the TRIANGLE DE LA PHYSIQUE FOUNDATION, and the ÉCOLE POLYTECHNIQUE for sponsoring this Seminar. Special thanks are due to CHANTAL DELONGEAS for the preparation of the manuscript.

BERTRAND DUPLANTIER, THOMAS C. HALSEY & VINCENT RIVASSEAU
Saclay, Houston and Orsay, May 2010

Glasses and Grains, 1–24
© 2011 Springer Basel AG

Glasses

Jorge Kurchan

Abstract. It is customary to present glasses as an outstanding unsolved question in condensed matter. The problem is the following: supercooled liquids in equilibrium appear to have typical relaxation timescales that diverge, as the temperature T is lowered, faster than with an exponential of $1/T$. Now, for particles with smooth, soft interactions, this may only happen as a consequence of a growing length of coherence between particle positions, diverging at $T = 0$. The mystery is that we do not actually see any recognizable form of order when we look at configurations.

Quite apart from this issue, there is another side to glasses, more relevant from the experimental point of view, the fact that in practice glasses are performing out of equilibrium dynamics: they are aging. Understanding, and even simply describing such a situation requires a new set of ideas and techniques.

1. Introduction

At school we are taught that heat is motion, and that constant molecular collisions are the explanation why a solid expands, and in general becomes more fluid as it gets hotter. When a system is cooled to sufficiently low temperatures, we thus expect it to collapse: molecules should become densely packed, leading to a form of matter that does not flow easily.

We are later surprised to learn that, in many cases, this molecular crowding is not a continuous process, but what happens rather is that upon cooling, all of a sudden the system arranges itself in an ordered manner, all particles spending most of their time around positions disposed regularly, in a periodic arrangement. Crystallisation is the first inherently collective phenomenon we become aware of.

There are however exceptions to this miracle of crystallisation, in which systems, upon cooling, seem to behave in the most naive manner, gradually becoming solid-like, with particles just moving slower and in a more constrained manner, but with no evident spatial order emerging: we then say that we have formed a glass.

This at first sight most unremarkable behaviour is, strangely enough, the one we understand the least.

The viscosity of a glass-former liquid (a substance able to avoid crystallisation) increases upon cooling without any important change in structure, but still in an explosive way: many orders of magnitude in only a few degrees Celsius. How are we to explain this, in the absence of anything sudden or remarkable happening to the arrangement of the particles? Having avoided the obvious miracle of ordering, glass formers present us with the mystery of their sudden change of behaviour, leaving us to wonder if there is a hidden-form organisation of matter, or an avoided "nearby" sharp transition, which we have yet to discover.

The problem in glasses, and why we consider it still open, is neither a question of fundamental interactions nor of practical calculations. On the fundamental side, we have plenty of models that exhibit a glass transition, and computers that can simulate by now very respectable times and sizes: they confirm that every microscopic element has already been put into the models. On the other hand, even if we do have limitations in our ability to compute things analytically, the situation is the same with liquids or dense gases, both subjects that are not usually described as a challenge. Our problem is instead one of interpretation: we are trained to believe that for every striking phenomenon there should be a set of ideas that is simple, invokes entities that have a clear definition, lends itself to a mathematical formulation, and is able to surprise us with a new prediction. We are only beginning to envisage such a theory.

2. Crystallisation

When we cool a liquid, crystallisation may occur all of a sudden. The energy then jumps to a lower value (Figure 1) – we say the system loses its latent heat – and from the microscopic point of view the system is now organised (Figure 2). The same situation arises with hard particles, with the volume and the inverse pressure playing the role of energy and temperature $(V, P^{-1}) \leftrightarrow (E, T)$.

A periodic distribution of matter has a spatial Fourier spectrum composed of delta contributions: these are the Bragg peaks (Figure 3). They are directly observable with diffraction measurements. Except at zero temperature, the instantaneous location of particles fluctuates around their truly ordered positions. For a crystal, these fluctuations do not affect the notion of order, since even in their presence there are Bragg peaks – and what is more, they pose no problem for our eye to recognise periodicity either.

Because we shall need to consider cases in which there is no periodicity, and no tool playing the role of a Fourier transform, it is convenient to detect order in an alternative fashion. The fact that there is an average density modulation can be directly seen from the fact that the time-average density (Figure 4)):

$$\bar{\rho}(x) = \tau^{-1} \int_0^\tau dt\, \rho(x,t) = \frac{1}{N\tau} \int_0^\tau dt\, \Sigma_a\, \delta[x_a(t) - x] \tag{1}$$

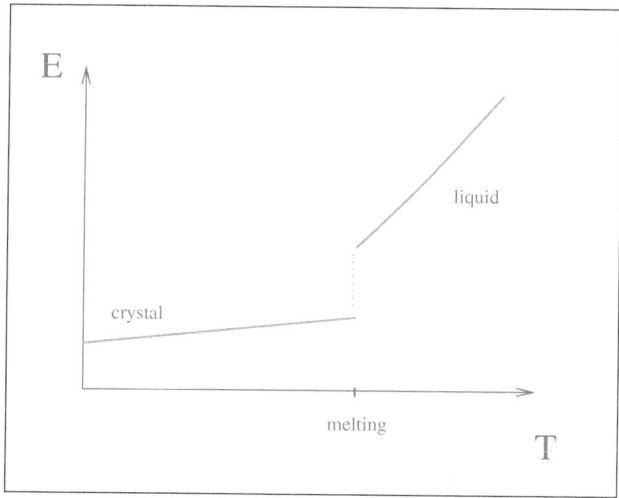

FIGURE 1. Energy versus temperature – or volume
versus $1/(\text{pressure})$.

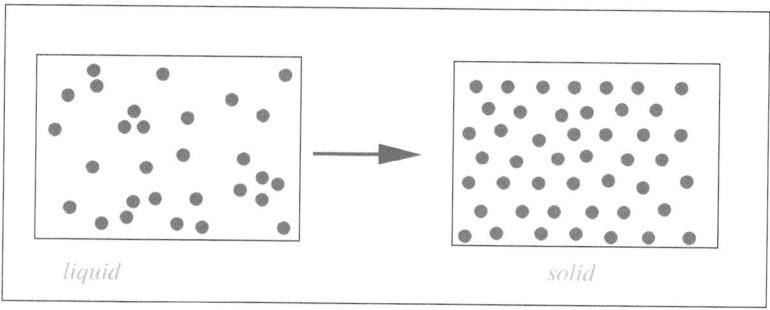

FIGURE 2. Ordering as the system jumps from a liquid (left) to
a crystalline configuration (right).

has a non-constant limit as $\tau \to \infty$ (taken *after* the thermodynamic limit). Another
useful way of conveying the same information is to consider a two-time autocorrelation function, as in Figure 4:

$$C(t, t_w) = V^{-1} \int dx \, [\rho(x, t)\rho(x, t_w) - \rho_o^2].\qquad(2)$$

In terms of $t - t_w$, there is a fast relaxation, corresponding to the rapid motion
including the phonons, but the correlation saturates to a plateau $C = V^{-1} \int dx$
$[\bar{\rho}(x) - \rho_o]^2$

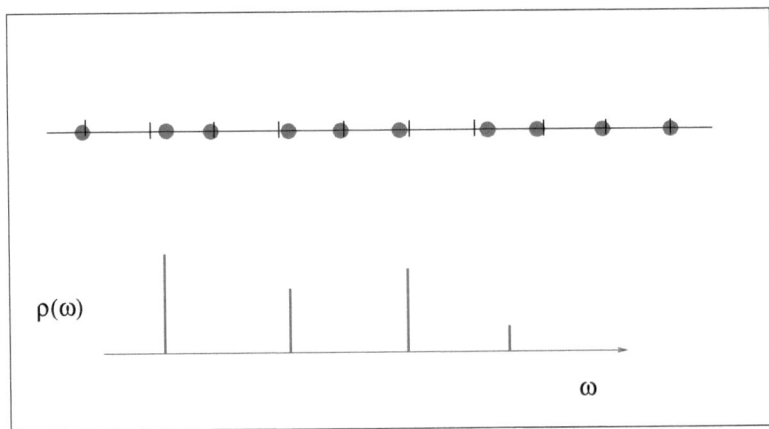

FIGURE 3. Periodicity, fluctuations and a Bragg peak.

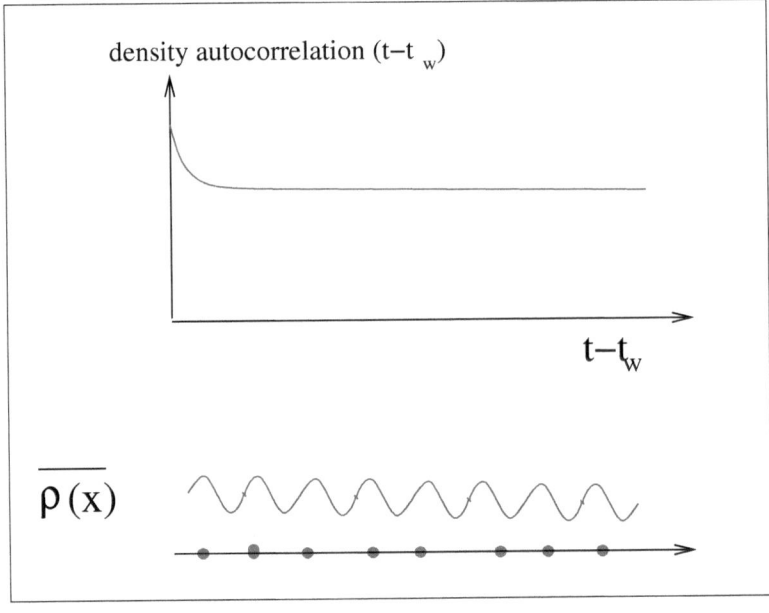

FIGURE 4. Autocorrelation function and time-averaged density
$\bar{\rho}(x) = \tau^{-1} \int_0^\tau dt\ \rho(x, t)$.

3. Collective nature of solidity: Arrhenius versus super-Arrhenius

The popular, generic term *jamming* as applied to solidification [2] may be sugges-
tive of rigidity arising from hard constituents in contact with one another, each
one blocking its neighbour. However, it is important to bear in mind that rigidity
is, at least for crystals and glasses, a collective phenomenon that does not require
hard constituents at all, and does not imply or require that any individual one be
blocked. The crystal example allows us to discuss in a very simple manner what being a solid does, and what it does not, mean. The property of having a permanent
(average) density modulation is one characteristic defining a solid. Another, more
explicit one, is the fact that they do not flow when subjected to an *infinitesimal*
stress [1, 29].

Consider first the case of soft particles (without a hard core) at finite temperatures, as in Figure 5. It is clear that any particle may exchange its position with

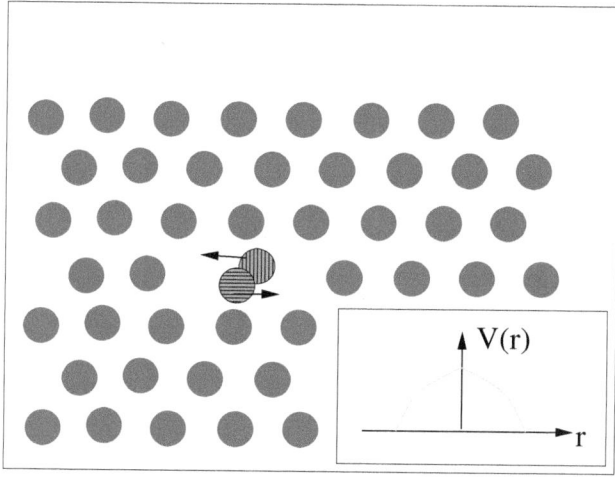

FIGURE 5. Permuting soft particles.

a neighbouring one with finite probability, so that there is no order in the particle
positions, if they are distinguished. Order is then a property of the density modulation, just as an army has permanent rank order independent of the changing
names of soldiers and generals. Another important point is that there can be no
order in a *finite* system, since for such a system there will be a finite probability
of being in any configuration, having started from any other. The same can be
said for a system of hard spheres (Figure 6), at finite pressure, because particles
can always "make way" for others to rearrange. And yet, we know that infinite
systems of this kind – soft spheres, hard particles at densities such that they do
not touch – do form solids in the thermodynamic limit.

Two further examples may be instructive. Consider the ferromagnetic Ising
model at $T = T_c/10$. Equilibrium is given by a state with positive and one with

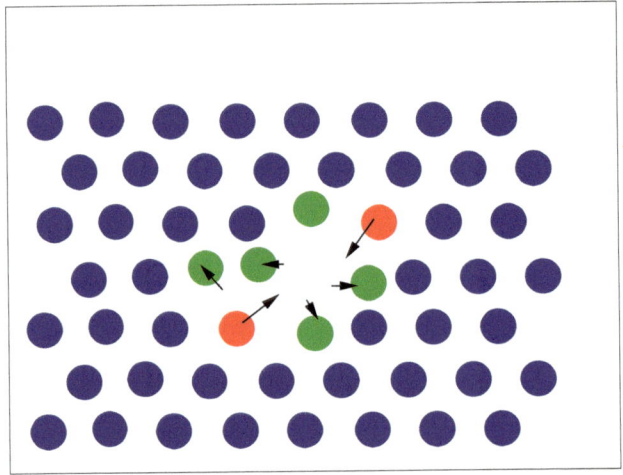

FIGURE 6. Permuting hard particles.

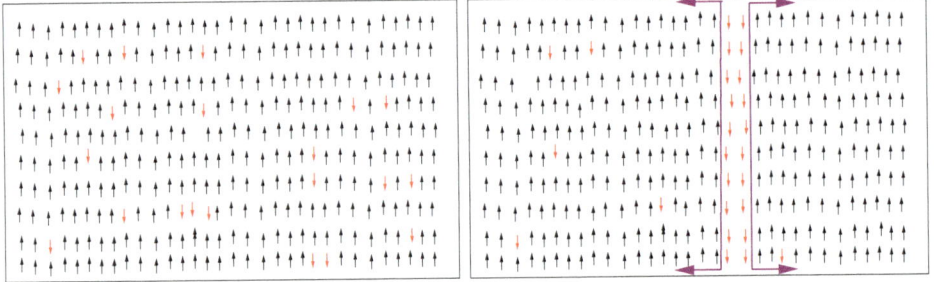

FIGURE 7. A collective, entropic, infinite barrier

negative magnetisation. The fact that an infinite system has a permanent magneti-sation, and that symmetry is broken, relies on the impossibility of the magnetisa-tion flipping. However, it is easy to find a path of constant energy leading from a typical configuration of the positive magnetisation state to a typical configuration of the negative magnetisation state. It suffices (Figure 7) to "herd" the minority down spins into a large stripe, and to grow this stripe laterally as a constant energy. The barrier is entropic in nature: it takes many simultaneous things to happen in order to assure the passage, and the probability of all of them occurring, though finite in a finite sample, becomes zero in the thermodynamic limit. Going back to the soft-sphere crystal, a spontaneous deformation like that of Figure 8, has an infinite energy barrier, because it involves an infinite amount of overlaps in the thermodynamic limit. Infinite entropic or energetic barriers are, in all these cases, collective phenomena.

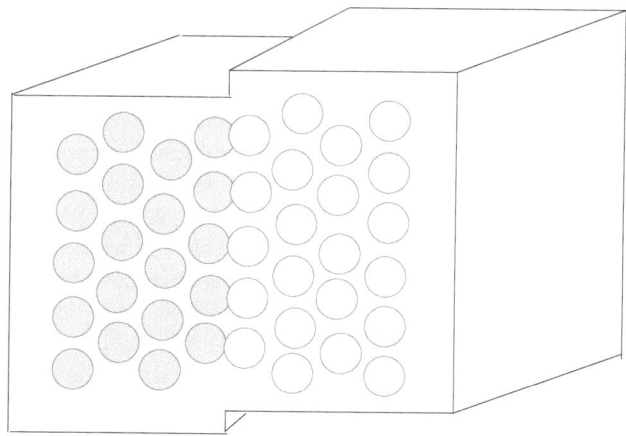

FIGURE 8. A collective, energetic, infinite barrier.

In contrast with the previous examples, one may have systems that are only solid because already its individual constituents are. In that case, even a finite version may be solid. Simple examples of this are depicted in Figure 9, where the spheres are assumed to be hard, or if they are not, the temperature is assumed to be zero.

A more subtle example of the same thing are the *kinetically constrained models* [3]. These are lattice models in which the particles have some forbidden

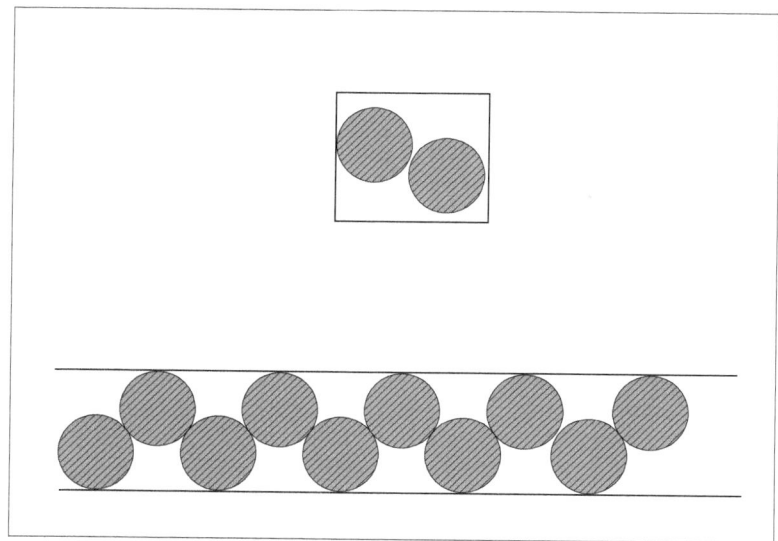

FIGURE 9. Non-collective rigidity.

moves. For example if their number of neighbours is higher than a certain number, the particle is immobile. The restrictions play, in this case, a role analogous to the hard constraints of Figure 9: as soon as they are partially lifted, infinite timescales and permanent modulations disappear.

In the situations in which rigidity and permanent modulations of density do not have a collective origin, for example in the case of *finite* systems, the timescales grow as $T \to 0$ or $P \to \infty$ in a typical activated (Arrhenius) manner. This is clear, because there is a finite barrier that takes more time to cross at lower temperatures. On the other hand, a collective system in the thermodynamic limit may have timescales that diverge at finite temperatures (e.g., the Ising model), or at any rate grow faster than with an exponential Arrhenius law. What we have just said can be made rigorous [4]: a system having a timescale that grows faster than exponentially *necessarily* has some equilibrium cooperativity length that diverges when the timescale diverges.

4. Avoiding crystallisation

Let us now turn to the situation when crystallisation does not happen. One can cool a liquid in such a way that the crystalline phase does not have the opportunity to nucleate. How easily this is done depends on the cooling protocol and on the nature of the liquid – a "good" glass-former is a poor crystal-former, and vice-versa.

The supercooled liquid just below the melting temperature is metastable, but in an innocent way: it can be considered to be in "local" equilibrium: if the temperature is not changed, the macroscopic observables do not evolve, and the equilibrium theorems (Fluctuation-Dissipation, Onsager reciprocity) hold. In other words, the supercooled liquid phase is in a situation similar to that of diamond, a mixture of oxygen and hydrogen at room temperature, or a current-carrying superconductor; which for all practical purposes ignore the possibility of nucleating graphite, water, or a lower supercurrent, and may be treated as equilibrium systems.

Upon cooling further, the viscosity grows dramatically, and the liquid reaches a point in which it falls out of equilibrium – but this time in a serious way. We can tell this because energy and viscosity now start depending on the cooling speed, and even if the temperature is held constant, they continue to evolve – as do all other macroscopic observables. This situation is completely unrelated to the existence of the crystal and very different from the innocent metastability of diamond or the oxygen-hydrogen mixture. The system is now in a situation in which something is constantly evolving so that, as we shall see, one can determine experimentally its "age" since it fell out of equilibrium and it became a glass.

Consider the cooling of a system as in Figure 10. For a fast cooling, the energy ceases to have its equilibrium value at a temperature T_2; for a slower process, this happens at a lower temperature T_1. We recognise the equilibrium

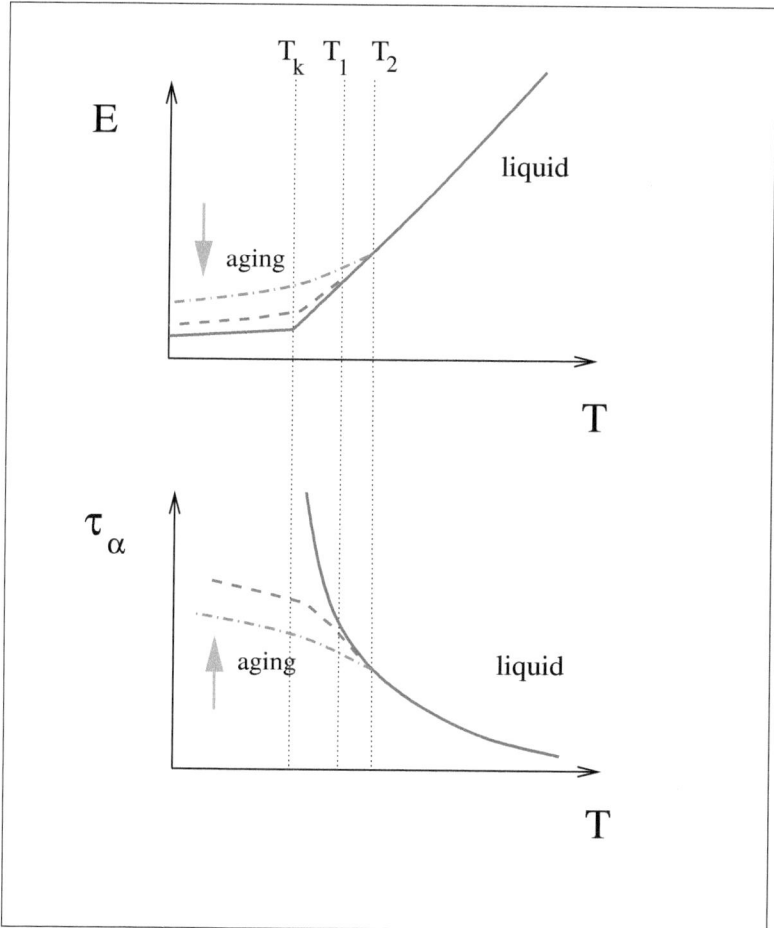

FIGURE 10. Different annealing speeds. The dashed lines indicate out of equilibrium situations, and are necessarily evolving in time. The full line is the result of infinitely slow cooling: energy has a nonanaliticity, and the relaxation timescale τ_α a divergence, if there is a true phase transition.

energy vs. temperature curve as the envelope beyond which all slower annealings coincide. What we have said about energies, can be said about the viscosities, or the relaxation times τ_α. Consider an autocorrelation function, for example (2). In the supercooled liquid phase, the autocorrelation falls in two steps: first to a plateau, and the second, in a much longer time τ_α, to zero. The first drop to the plateau is analogous to the one observed in a crystal (Figure 4), and is a consequence of rapid vibrations, while the second drop – entirely absent in a crystal

J. Kurchan

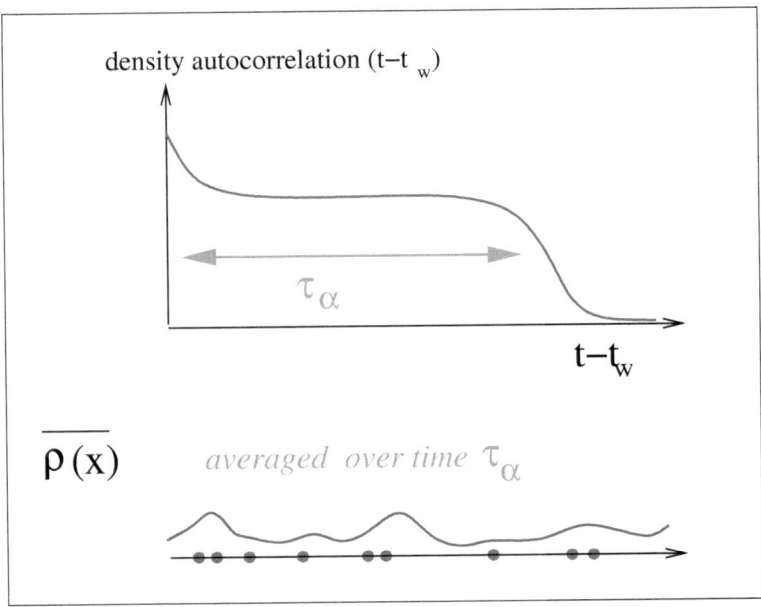

FIGURE 11. Density profile in an α scale.

– reflects the large rearrangements that a liquid can afford to make. A direct way to picture the α relaxation is to consider, as in Figure 11, the analogue of Figure 4: averaging out the rapid vibrations, as we did in the crystalline case, we obtain an amorphous density profile that does not last forever, but takes a time $\sim \tau_\alpha$ to evolve. Within (metastable) equilibrium, τ_α increases rapidly as the temperature is decreased, reflecting the increase in the viscosity (Figure 12, right). When the system is further cooled and falls out of equilibrium, the autocorrelation function is no longer an exclusive function of temperature, but depends also on history via the "waiting" time t_w (Figure 12, left): in particular, the system needs time to become more viscous. This is the "aging" phenomenon. Still, at a temperature T_1 (Figure 12, top), eventually τ_α reaches its equilibrium value, although this may take so long that we only observe aging.

This is how glasses present themselves to us in real, experimental life. We may still be curious to know whether there is a temperature T_K below which aging lasts forever, equilibrium is never achieved, and the timescale τ_α becomes infinite. If this were the case, one could ideally consider samples with a permanent, amorphous, averaged density profile $\bar{\rho}(x)$, a solid just like a crystal in all but the spatial periodicity. The discussion above about collective rigidity implies that if such states with permanent spatial modulation of density exist at finite temperature, then necessarily they involve a coherent behaviour of particles that only exists rigorously in the thermodynamic limit, and requires the divergence of

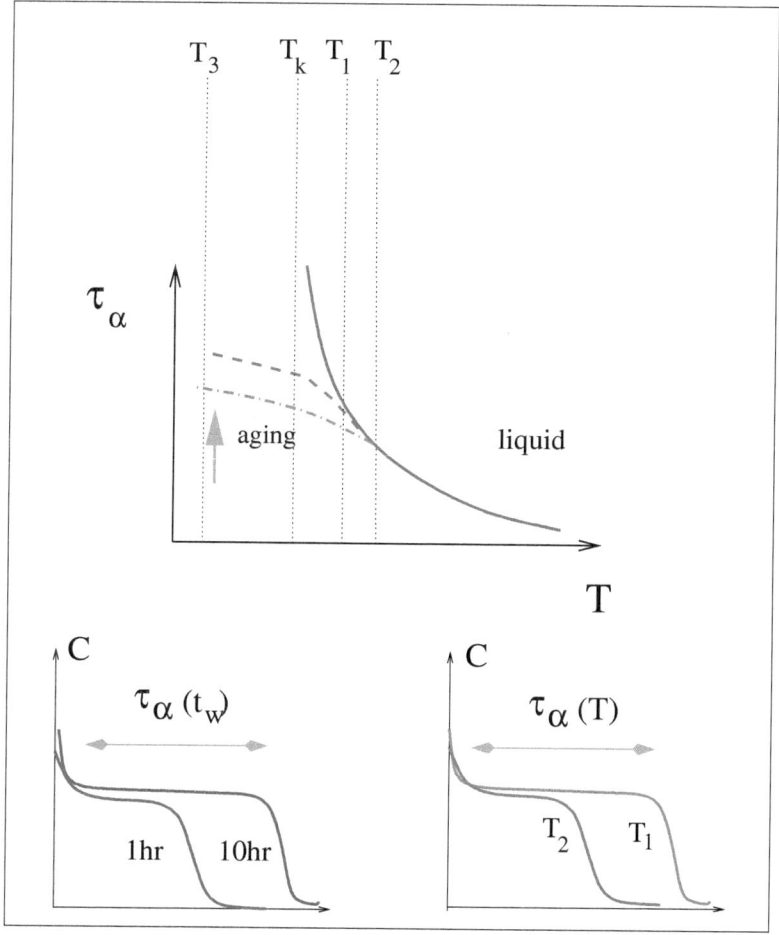

FIGURE 12. α time relaxation depends on temperature in equilibrium, and on the waiting time out of equilibrium.

some correlation length. This is even the case if $T_K = 0$, but τ_α grows faster than an Arrhenius law $\tau_\alpha \sim e^{b/T}$ [5].

Within an α scale, we can classify configurations as in Figure 13: two configurations are considered to be in the same metastable state if the density profiles obtained starting from either one, and averaging over a time τ_α, coincide up to, say, $\tau_\alpha^{-1/2}$ (i.e., within the statistical error). This is sometimes depicted in a "landscape" picture (Figure 13, right). All the configurations that yield the same profile constitute a "state", and their number yields the entropy within the state. More importantly, the logarithm of the number of states (per unit volume) is by definition the *complexity* Σ [6].

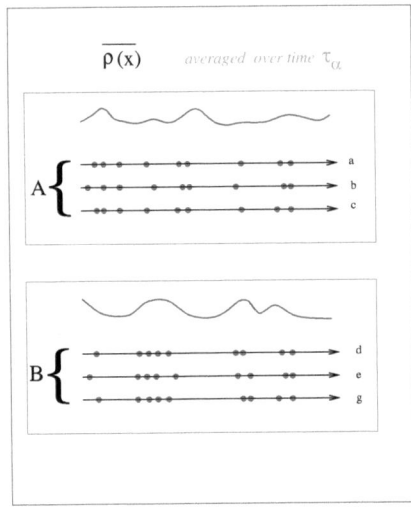
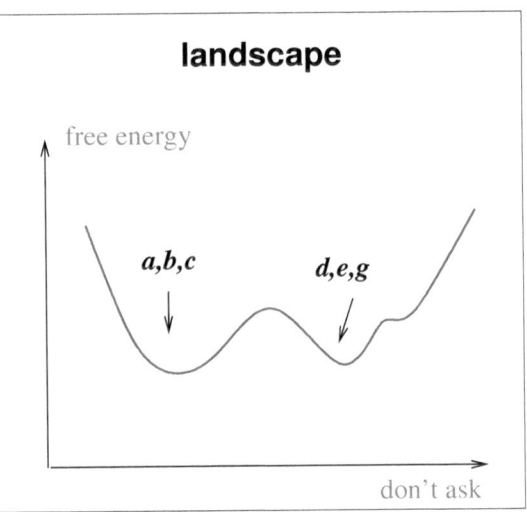

FIGURE 13. Free Energy Landscape picture.

5. Short digression: the nucleation argument

Two phases

Before continuing, it is useful to recall the nucleation argument, which allows us to conclude that for finite-dimensional systems at non-zero temperature, with short range interactions and soft potentials – these are all we consider here – a phase with a free energy density higher than the equilibrium one *cannot be stable*. This means that true stable states have a free energy that exceeds the equilibrium one at most by a *subextensive* amount.

One considers phases **a** and **b** with free energy densities $f_a > f_b$. In the phase **a**, a droplet of radius r of the phase **b** costs at most a surface energy σr^{d-1}, with $\sigma \leq 0$, and involves a gain $-(f_a - f_b)r^d$. In terms of r, we have:

$$\delta f(r) = \sigma r^{d-1} - (f_a - f_b)r^d \qquad (3)$$

which has a maximum $\delta f(r^*)$ at a critical radius r^*:

$$r^* = \frac{(d-1)\sigma}{d(f_a - f_b)} \quad \rightarrow \quad f(r^*) \propto \frac{\sigma^d}{(f_a - f_b)^{d-1}}. \qquad (4)$$

The droplet growth is activated up to r^*, with Arrhenius probability $\sim e^{-\delta f(r^*)/T}$, and then proceeds downhill until phase **b** prevails. We have found a path leading from state **a** to phase **b** with a finite free energy barrier requiring a finite number of moves: it is perhaps not the best path, but it gives an upper bound on the probability of nucleation. The only way in which the droplet will not unstabilise **a** is that either $\sigma = \infty$ (which requires hard, or long-range interactions), or that $(f_a - f_b) \rightarrow 0$ as $N \rightarrow \infty$.

The nucleation argument is stronger than this: it implies that state **a** cannot have *any* sub-region of extensive volume having a free energy density larger than the corresponding one of state **b**.

Entropic drive

A situation that arises in supercooled liquids is that a system has many options of phases for nucleating. The question then is: does this multiplicity increase the probability of nucleation? The argument against says that it does not, since once one is nucleating one phase, in what way does it help to know that there was another option? Or, put in another way, how can the system know, when it is going somewhere, that there are other options out there?

To clarify the point, it is best to do a small calculation. Consider a system at very low temperature, activating its escape out of the spherical crater $V(r)$ in Figure 14. Starting from a spherical distribution concentrated at the bottom,

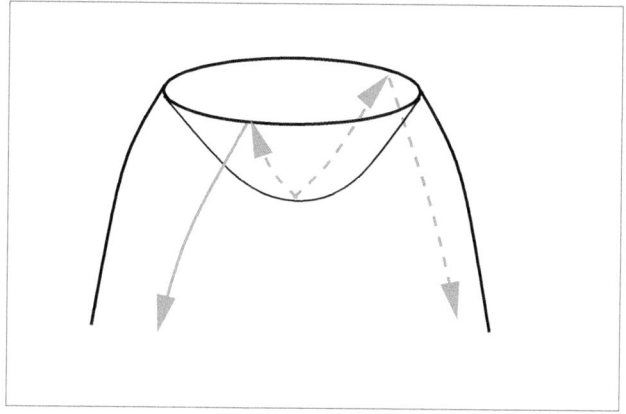

FIGURE 14. Two trajectories escaping a crater.

the particle follows (say) a Langevin process, and the probability evolves via a Fokker-Planck equation.

$$\dot{P} = \nabla \left[T\nabla + \nabla V \right] P. \tag{5}$$

Assuming the distribution was spherically symmetric at the start, it will remain so, and we may go to spherical coordinates:

$$\dot{P} = \frac{1}{r^{d-1}} \left[T \frac{d}{dr} \left(r^{d-1} \frac{dP}{dr} \right) + P \frac{d}{dr} \left(r^{d-1} \frac{dV}{dr} \right) + \frac{dV}{dr} \frac{dP}{dr} \right]. \tag{6}$$

Putting $\tilde{P} \equiv r^{d-1} P$ we get the radial diffusion equation

$$\dot{\tilde{P}} = \left[T \frac{d^2}{dr^2} + \frac{d}{dr} \left(V(r) - T(d-1) \ln r \right) \right] \tilde{P}. \tag{7}$$

This is the dynamics in a potential V corrected by precisely the entropy(the logarithm of the volume) $(d-1)\ln r$ of a shell of radius r. Indeed, the different possibilities do add up, and help in lowering the effective barrier. With hindsight, we can justify this even at very low temperatures by noting that before a passage is actually made, many attempts that barely failed have been made – and these take all possible paths. We shall use this in what follows.

6. Configurational entropy and metastable states

Local mean-field

Landau theory consists of writing a free energy in terms of a space-dependent order parameter. This free energy functional contains an entropic term that takes into account all the rapid thermal fluctuations, and temperature enters only as a parameter. The order parameter itself represents the time-average of the microscopic variables, for example the magnetisation is the time-average of the spins. For simple forms of order, although we know that the theory is not exact, and in general leads to the wrong exponents, it gives a satisfactory qualitative picture. Phase transitions appear when the minima of the free energy functional are a set of symmetry-breaking solutions related between one another by the symmetry operation.

In glassy systems, when we attempt such a mean-field approach, for example the Thouless-Anderson-Palmer (TAP [7]) approach to spin glasses, we find that at low temperatures the free energy functional now has an exponential number of solutions, rather than two as in a ferromagnet. For the case of a liquid, the analogue of the local magnetisation is clearly our time-averaged density $\bar{\rho}(x)$, and a closely related approach is the so-called density functional theory. We are given a free energy functional in d-dimensional space [8]:

$$F[\rho(\mathbf{x})]$$
$$= \int d^d\mathbf{x}\ \rho[\ln\rho(\mathbf{x}) - 1] - \frac{1}{2}\int d^d\mathbf{x}\,d^d\mathbf{x}'\ [\rho(\mathbf{x}) - \rho_o]C(\mathbf{x} - \mathbf{x}')[\rho(\mathbf{x}') - \rho_o]. \quad (8)$$

Here $C(\mathbf{x} - \mathbf{x}', \rho_o)$ is the liquid direct correlation function at average density ρ_o computed within some, such as the Percus-Yevick, approximation. For short range interactions, $C(x)$ is short ranged. We look for the "local" free energy minima that satisfy:

$$\frac{\delta F[\rho(\mathbf{x})]}{\delta\mathbf{x}} = \ln\rho(\mathbf{x}) - \int d^d\mathbf{x}'\ C(\mathbf{x} - \mathbf{x}', \rho_o)[\rho(\mathbf{x}') - \rho_o] = 0. \quad (9)$$

At low average densities ρ_o, the spatially constant "liquid" solution dominates. As the density increases, a periodic, "crystalline" solution appears. What is interesting from the glassy point of view [11], is that in the high density regime, there appear also many non-periodic "amorphous" solutions, as depicted schematically in Figure 15. Each one of these is supposed to represent a metastable glassy state, as described in the previous section. These states are local minima of (8) satisfying (9).

Now, as we have seen, the nucleation argument implies that as soon as we go beyond the mean-field approximation and add fluctuations to this picture, solutions with free-energy density $O(1)$ above the lowest are unstabilized. We already know that if the crystal has lower free energy, everything is metastable with respect to it, but we have argued that we could ignore this. Here we are saying that in fact essentially all solutions schematised in Figure 15 are metastable *even if*

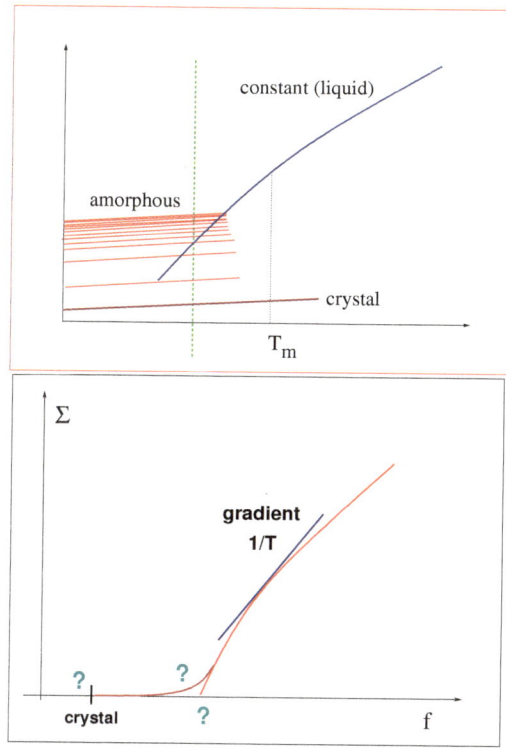

FIGURE 15. Complexity Σ versus free energy f.

we neglect the crystal, they will nucleate one onto the other and only the lowest of the amorphous ones are truly stable – or only unstable with respect to crystal nucleation (which takes an altogether different scale).

We are now embarrassed: we have claimed that the amorphous solutions of (9) represent a glass, but in fact, since all but the very lowest are metastable: *they correspond to the liquid phase*. Worse of all, we have now *two different* representations of the liquid phase, one as a constant solution, and one as a set of metastable amorphous solutions.

A sum rule

Let us be more precise: consider all amorphous solutions at temperature T, summed with the Boltzmann weight:

$$Z = \Sigma_{solutions} \quad e^{V[\Sigma(f) - \beta f]}. \tag{10}$$

This sum is dominated by the saddle point, yielding:

$$\frac{d\Sigma}{df} = \frac{1}{T}. \tag{11}$$

The solution of this equation is obtained with a tangent construction as in Figure 15 (right). For sufficiently high temperatures, the saddle-point free energy corresponds to solutions that are well above the lowest, so that the Boltzmann weight is dominated by an exponential number of metastable states with a finite lifetime. The question is now: what is the true representative of the supercooled liquid, these states or the constant solution? The answer is quite surprising: it turns out that within the models for which a full solution is available (more about these later), there is a range of temperatures where *both results coincide*, so that the liquid is given *twice*, once by a constant profile and once as a packet of amorphous solutions. The latter give us the metastable states characteristics, such as one observes in that regime within the α scale. This sum rule has not been, to the best of my knowledge, discussed or tested within these approaches "with space".

An objection may now arise: given that states that dominate in the liquid phase have the same free energy, how do we know that the free-energy barrier separating is not infinite? This is indeed a valid question, since our nucleation argument only showed that barriers are finite between states with a finite free-energy density difference. Here we have to invoke the entropic drive we mentioned above: just like in the escape from a crater, the system has many nucleation paths (roughly $e^{\Sigma r^3}$) leading to different density profiles, and this modifies accordingly the activation time, cf. Eq. (7).

The transition

What happens within this approximation when we lower the temperature? Just looking at Figure 15 (right), we see that if the Σ versus f curve reaches zero with a finite gradient equal to, say, β_c, then below $T_K = 1/\beta_c$ the equilibrium distribution freezes in the lowest amorphous states. These are the ones that *are stable* (except for crystal nucleation), and they represent the true glass phase. Hence, we have obtained the glass transition as a condensation into a handful of low-lying density profiles, coming from a supercooled liquid made of similar, though metastable, states representing the liquid. This is the Kauzmann scenario [9].

If, on the contrary, $\Sigma(f)$ reaches zero with infinite slope (a possibility advocated in [10], we have that the glass transition temperature is zero. Note again, that this will not make it more trivial, since the vanishing of entropy, even at $T = 0$, implies some form of order. We cannot exclude yet other possibilities, and the question marks on Figure 15 are there to express this.

Reading the complexity and a coherence length from $\bar{\rho}$

At any rate, it is interesting to note that as we find deeper and deeper amorphous states, we expect that a correlation (or coherence) length will grow. A concrete realisation of this length is the following [18]: given an infinite system, we choose a block of size ℓ, and see how far we have to go in order to find – within a certain precision – a block with the same configuration of $\bar{\rho}$. This length is exponential in ℓ^d in a truly random density configuration, but will be subexponential in a crystal, a quasicrystal, and more general objects with hidden forms of order. The distance of patch-repetition gives a direct measure of the complexity: if a patch repeats every $e^{\ell^d/\Sigma}$, then Σ is the complexity. Note that $\Sigma \to 0$ implies a diverging length.

7. Analogy with – and lessons from – chaotic systems

The equations (9) are analogous to the equations of motion of a dynamical system, with space playing the role of (multidimensional) time. A constant "liquid" solution is then analogous to a stationary point, a crystal to a periodic solution, and amorphous solutions correspond to chaotic orbits. This similarity between dynamical systems that are chaotic in time, and glassy systems that are chaotic in space, was pointed out many years ago by Ruelle [15]. As it stands, the analogy is not perfect, since we demand not only that the density profile be a solution of (9), but that in addition it be a deep minimum of (8). In order to make the analogy closer, we may consider a dynamical system, in which in addition we look for minimal solutions of the action

$$S = \int dt \, L(q, \dot{q}) \tag{12}$$

so that (12) plays the role of (8), and the (Lagrange) equations of motion

$$\frac{\delta S}{\delta q(t)} = 0 \tag{13}$$

play the role of (9). A realisation of this appeared in the theory of charge-density waves [16, 17], in particular in the Frenkel-Kontorova model, for which the local energy minima of the model are given by the trajectories of the "standard map", which has both regular and chaotic orbits.

In order that the action plays the role of a free energy, we need it to be bounded from below. This is not in general the case, and one needs for example that the potential be bounded from above. This should not worry us: in fact, one can take Lagrange's equations (13) as the analogue of (9) and any functional $\mathcal{A} \equiv \int dt \, A(q, \dot{q})$ as the analogue of (8). One is then computing trajectories that are a large deviation of $\mathcal{A}$ [13].

What seems to happen [16, 17, 13] when we look for trajectories that are solutions of the equations of motion of a *chaotic* system *and* minimise globally some quantity $(S, \mathcal{A})$ is that the trajectories that dominate are periodic or quasiperiodic, or have in general some form of regularity. In such systems, these trajectories are *not* in a regular region of phase-space, they are unstable and buried in the middle of the chaotic sea.

With this analogy in hand, we now consider again the solutions of (13), but classifying them according to (12) (when possible), or with any other functional $\mathcal{A}$, and the correspondence is:

- Stationary points correspond to the liquid solution.
- Periodic orbits in regular (unstable) regions correspond to a crystal.
- Chaotic trajectories correspond in general to the supercooled liquid phase.
- The glass state corresponds to chaotic solutions that minimise the chosen functional (S or $\mathcal{A}$). These may be *unstable* periodic or quasiperiodic orbits [13], or have a more subtle form of order [18].

  However, note that *even when they are periodic, these extremal orbits are very different from an orbit of a regular system*, in that they are in the middle of a sea of chaotic solutions, and are dynamically unstable in the sense that a perturbation in the boundaries will change the orbit dramatically.

  An orbit minimising $\mathcal{A}$, but with arbitrary boundary conditions in the coordinates, will approach the unstable periodic one, shadow it for most of the time, and then go to the prescribed endpoint.

If we take this analogy seriously, the ideal glass state may well be spatially ordered (periodically, quasiperiodically, or in general with frequent motif repetition), but it would be of a different kind than a crystal or quasicrystal: the density profile would be surrounded by disordered solutions, just as the unstable periodic orbits which exist in purely chaotic systems, intermixed with the chaotic orbits as the rationals are with the reals.

8. Glasses in the real world: aging

In the real world, glasses know nothing about an ideal transition, they are just systems slowly working their way to equilibrium, insensitive to whether such an equilibrium is eventually reachable or not. It would seem that the phenomenology of such a situation would be all but universal, and that a theory of such a situation is hopeless. This turns out not to be the case.

As mentioned above, in the aging phase the α relaxation time increases with time, as does the viscosity [1]. When stress is applied to a plastic bar below the glass transition, the contraction happens in two steps: a fast elastic step followed by a slow "creep" motion [14]. Figure 16 shows the classical experiments by Struik, where the creep motion as a function of time is measured for a sample at different "waiting" times after it was quenched below the glass transition. Remarkably, the creep step takes a time roughly proportional to the waiting time, and this in a range from minutes to years. Clearly, no equilibrium theory can explain this behaviour,

---

[1]Note that this would happen also in an imperfect crystal which is gradually healing its defects.

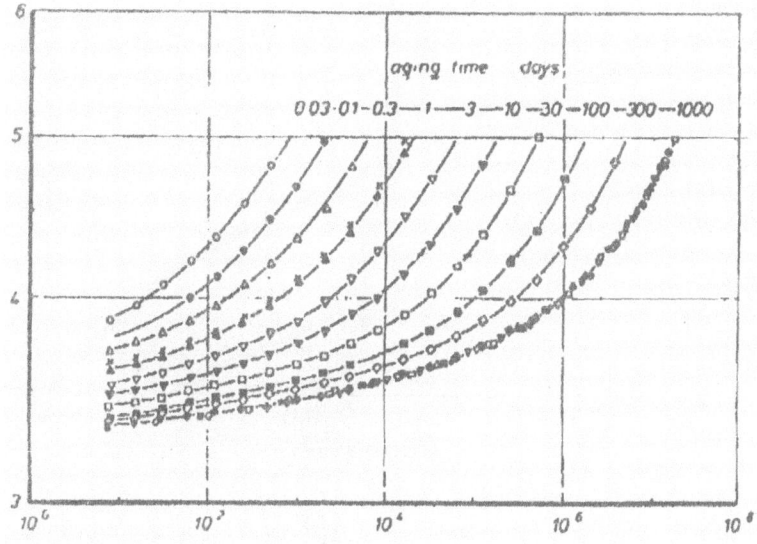

FIGURE 16. Struik's classical experiment: Evolution of creep after a stress applied at different waiting times.

which has been obtained in a variety of glassy systems: plastics, colloids, spin glasses, etc.

These experiments concern a *response* to a field. Similar curves are obtained when one considers a *correlation*. In equilibrium, these correlate by the fluctuation-dissipation relation, which states that the response of an average value $\chi(t, t_w) = \frac{\delta}{\delta\langle X\rangle(t)}$ to a field that acts on Y from time t_w, is given in terms of the correlations as:

$$T\chi(t, t_w) = \langle X(t)Y(t)\rangle - \langle X(t)Y(t_w)\rangle = C(t, t) - C(t, t_w). \tag{14}$$

In equilibrium, a χ versus C plot, parametric plot of all (t, t_w) should yield a straight line with slope $-1/T$.

A very different thing happens when we consider [22] the correlation and response of an aging glass (Figure 17). All points fall on a line, which is now composed of two apparently straight segments. For t close to t_w, corresponding to high frequencies, one obtains a line with gradient $-1/T$ as in equilibrium, but for t and t_w farther apart – precisely in the range where the response is the creep motion – one obtains a different line with slope, say, $-1/T_{eff}$. The remarkable fact is that T_{eff} is the same (for the same time regime), for many different observables, suggesting that T_{eff} is a genuine temperature. Indeed, one can show that this is what a thermometer coupled to the slow degrees of freedom would measure [22]. This way of approaching the effective temperature comes to us from the analytic solution of the aging dynamics of the random first-order theory (see below), but

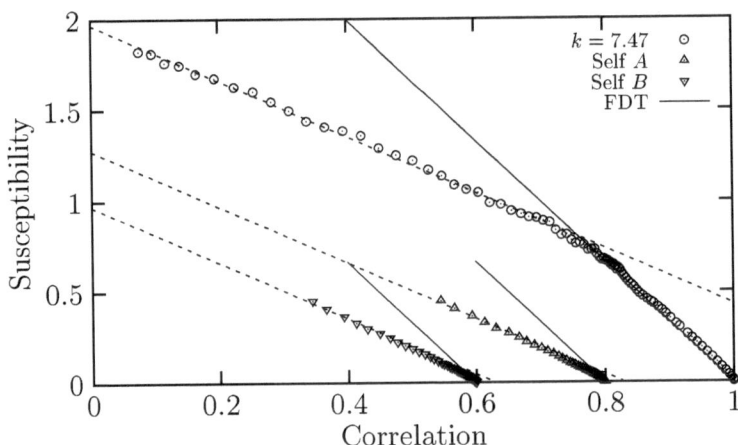

FIGURE 17. Effective temperatures: Response versus correlation for a binary mixture glass. The three lines correspond to density autocorrelations, and to diffusion versus mobility for each kind of particle. The autocorrelation curve shows the two-temperature behaviour, while the diffusion only the effective temperature, since it is a low-frequency quantity. The effective temperatures seem to coincide, as witnessed by the fact that the segments are parallel. Taken from Berthier and Barrat [23].

it seems to be the same kind of idea proposed at a phenomenological level many years ago by Tool [24].

One should beware of simplistic explanations: what is most important to keep in mind is that these effective temperatures are *not* due to some structure that has remained frozen at the configuration it had when the system crossed the glass temperature: since that time the system has decorrelated substantially, and the identity of the particles responsible for the aging motion and the effective temperature *is constantly changing.*

9. Random first-order theory

Random first-order theory is, or starts from, a family of models that are asserted to be for glasses what the Curie-Weiss (fully-connected) models are for a ferromagnet. First, it was observed in [31] that spin models with random disorder of the form

$$E = \sum_{ijk} J_{ijk} s_i s_j s_k \tag{15}$$

with J_{ijk} random interactions, reproduce some of the phenomenological features of glasses. Models like this have a static transition like Derrida's Random Energy

Model (REM) [35]. The mechanism is like the one described in Section 6, where the measure freezes at a certain T_K: this is indeed Kauzmann's scenario [9] for an ideal glass transition, with random energy levels playing the role of states.

Next, one can study the relaxational dynamics with this energy function. Remarkably, in the high temperature phase, the dynamics turn out to be exactly described by the Mode-Coupling (MCT) equations [33], which are a widely studied model of the first stages of approach to the glass transition from the liquid side. There is a Mode Coupling Transition at a certain $T_d > T_K$, known to be an artifact of the approximation, and within the present perspective one can understand easily why the MCT transition has to go away in finite dimensions.

Between T_d and T_K the *sum rule* mentioned in Section 6 is strictly obeyed: the description of the liquid state may be made in terms of many metastable glassy states, or a single high-temperature one, and both descriptions strictly coincide thermodynamically.

You are not forced to restrict the dynamics to the high temperature phase. If you quench the models to low temperatures, it turns out that the system does not equilibrate: it "ages", just like true glasses [21]. When one analyses the properties of observables out of equilibrium, one discovers [22] that the slow fluctuations behave as if they were "thermalized" in an effective temperature T_{eff}. As mentioned above, "fictive temperatures" have been around since the 1940s [24], and it is likely that what one has discovered is precisely a non-phenomenological version of that. One may also study how the system responds to forces that do work on it: one finds the generic phenomenology of "shear thinning" of supercooled liquids, and in some cases you can explain the much more rare "shear thickening" of certain glasses.

One may go back and study the free energy landscape, defined by the TAP [7] equations, something that was not available in a pure mode coupling context. One recovers the main features (importance and location of saddles, marginality, etc) that were discussed many years ago by Goldstein [36] at a phenomenological level.

The next extension of the random first-order scenario, is getting rid of the quenched disorder (the J_{ijk}), which are artificial and foreign to the problem. This has been done successfully, and by the 1990s there was a plethora of fully connected models having the same properties as (15). As in all mean-field situations, we wish to "put some space" into the formalism, in this way getting a Landau theory which, although we know will not capture fully the essence of finite dimensions, will at least give us a first hint. This was done by Mézard and Parisi [32] within the hypernetted chain and other approximations, and using the replica trick [30]. Approaching the mean-field-with-space approximation with a free-energy TAP [7] rather than a replica formalism, should finally give us a systematic and well-controlled way to go back to a density functional formalism like the one described in section 6 – which we now recognise as a form of the "random first-order" scenario.

As we have seen in section 6, the next big question is how to include fluctuations beyond a mean-field, which will inevitably unstabilise metastable solutions, and re-express the liquid in terms of those. This has been argued phenomenologically in the so-called *mosaic picture* [37, 38], with a degree of success [39]. A line that has not, to the best of my knowledge, been followed by many is an analytical study of the constraints of a theory with space. How does one define rigorously the complexity of density profiles, in analogy with the Kolmogorov-Sinai entropy? Are the lowest free energy solutions regular, and what is the correlation length that defines them (see discussion in [18])? What is the relation between configurations of a crystal with defects and the lowest amorphous ones; do they merge one into the other?

10. Is mean-field circumstantial or essential?

Let us recap. We start out trying to explain why it is that a liquid may become essentially solid by changing the temperature by a few degrees, with barely any detectable change in its structure.

Clearly, the question whether there is a diverging timescale, or rather, whether the longest (α) timescale is as long as it could be – that is, equal to the time of nucleation of a crystal and not shorter – is one which we may as without an approximation scheme in mind. However, in attempting an explanation we introduce notions such as metastable state, complexity, mosaic, effective temperature. We are limited in our analytic powers, and we resort to mean-field like approximations, or diagrammatic resummations[2] in order to obtain results.

The question we may ask is whether the concepts themselves are inherently mean-field in nature. Clearly, this is the case of finite free-energy density metastable states, and hence the complexity: once we step outside mean-field we need to specify a lifetime above which we call a state a state. Similarly, mosaics carrying a state label which has a meaning locally in space (rather than globally for the whole system) are also mean-field constructs, and so on. Even the definition of "activated" processes is also related to an approximation, since at the end of the day they are defined as being non-analytic corrections in the mean-field parameter.

If it turned out that our mental constructs are inherently "mean-fieldy", this could pose a problem in cases that are far removed from being exactly of that kind, but they could still provide the best (approximate) approach to thinking of the glass transition. This situation would not be without parallel in other branches of physics: for example superconductivity [26], superfluidity [27], elasticity [28] and rigidity [29] are in principle, but not in practice, undermined by activation.

---

[2]I assimilate diagrammatic resummations with mean-field treatments because one can always find a model for which the resummation is exact, and can be thought of as some form of mean-field disordered system [25]

References

[1] P.M. Chaikin and T.C. Lubensky, *Principles of Condensed Matter Physics*, Cambridge Univ. Press 1995.

[2] Andrea J. Liu and Sidney R. Nagel, *Jamming and Rheology Constrained Dynamics on Microscopic and Macroscopic Scales*, Taylor and Francis, (2001).

[3] G.H. Fredrickson and H.C. Anderson, Phys. Rev. Lett. **53** (1984), 1244. F. Ritort and P. Sollich, Advances in Physics **52** (2003), 219.

[4] A. Montanari and G. Semerjian, J. Stat. Phys. **125** (2006), 23.

[5] Collective behaviour and a thermodynamic phase transition at $T = 0$ would be implied if the law that hold turns out to be the one proposed in: YS. Elmatad, D. Chandler, JP. Garrahan, J. Phys. Chem. B **113** (2009), 5563.

[6] Counting a continuous set of profiles is not in itself problematic: the same situation arises when we count the entropy of trajectories in classical mechanics, and there are well established definitions involving coarse graining of space to do this unambiguously, see, e.g., G. Paladin, A. Vulpiani, J. of Phys. A: Math. and Gen. **19** (1986), L997.

[7] D.J. Thouless, P.W. Anderson and R.G. Palmer, Phil. Mag. **35** (1977), 593.

[8] T.V. Ramakrishnan and M. Yussouff, Phys. Rev. B **19** (1979), 2775.

[9] A.W. Kauzmann, Chem. Rev. **43** (1948), 219.

[10] F.H. Stillinger, J. Chem. Phys. **88** (1988), 7818.

[11] Y. Singh, J.P. Stoessel, P.G. Wolynes, Phys. Rev. Lett. **54** (1985), 1059;
C. Dasgupta, Europhys. Lett. **20** (1992), 131;
C. Dasgupta and S. Ramaswamy, Physica A **186** (1992), 314.

[12] J. Kurchan and D. Levine, arXiv:0904.4850.

[13] Khanh-Dang Nguyen Thu Lam, Jorge Kurchan, Dov Levine, J. Stat. Phys., to be published, arXiv:0907.1807.

[14] L.C.E. Struik, Physical aging in amorphous polymers and other materials, Elsevier Scientific, Amsterdam; New York (1978).

[15] D. Ruelle, Physica **A 113** (1982), 619–623.

[16] S. Aubry, Physica **7D** (1983), 240, J. Physique **44** (1983), 147.

[17] S.N. Coppersmith and DS. Fisher, Phys. Rev. **B28** (1983), 2566.

[18] J. Kurchan and Dov Levine, *Order in glassy systems*, arXiv:1008.4068.

[19] J. Hubbard, Phys. Rev. B **17** (1978), 494.

[20] S. Aubry, J. of Phys. C: Solid State Phys. **16** (1983), 2497.

[21] L.F. Cugliandolo, J. Kurchan, Phys. Rev. Lett. **71** (1993), 173.

[22] L.F. Cugliandolo, J. Kurchan, L. Peliti, Phys. Rev. **E 55** (1997), 3898.

[23] L. Berthier and J.L. Barrat, J. Chem. Phys. **116**, 6228.

[24] A.Q. Tool, J. Am. Ceram. Soc. **29** (1946), 240.

[25] R.H. Kraichnan, J. Math.Phys. **3** (1962), 475, *ibid.* (1962) 496.

[26] J.S. Langer and V. Ambegaokar, Phys. Rev. **164** (1967), 498; J.S. Langer and M.E. Fisher, Phys. Rev. Lett. **19** (1967), 560.

[27] J.S. Langer and M.E. Fisher, Phys. Rev. Lett. **19** (1967), 560.

[28] A. Buchel and J.P. Sethna, Phys. rev. lett. **77** (1996), 1520.

[29] F. Sausset, G. Biroli and J. Kurchan, *Do solids flow?*, Journal of Statistical Physics **140** Nr. 4 (2010), 718–727.

[30] M. Mézard, G. Parisi and M.A. Virasoro, *Spin-Glass theory and beyond* (World Scientific, Singapore, 1987).

[31] T.R. Kirkpatrick and P.G. Wolynes, Phys. Rev. A **35** (1987), 3072,
     T.R. Kirkpatrick and P.G. Wolynes, Phys. Rev. B **36** (1987), 8552.
     T.R. Kirkpatrick and D. Thirumalai, Phys. Rev. B **37** (1988), 5342.

[32] M. Mézard and G. Parisi, J. Phys. A: Math. Gen. **29** (1996), 6515,
     M. Mézard and G. Parisi, Phys. Rev. Lett. **82** (1999), 747.

[33] W. Götze, *Liquids, Freezing and the Glass Transition*, edited by J.P. Hansen, D. Levesque, J. Zinn-Justin, Les Houches. Session LI, 1989 (North-Holland, Amsterdam, 1991), 287.

[34] E.-J. Donth, *The Glass Transition: Relaxation Dynamics in Liquids and Disordered Materials* (Springer, 2001).

[35] B. Derrida, Phys. Rev. B **24** (1981), 2613
     D.J. Gross and M. Mézard, Nucl. Phys. B **240** (1984), 431.

[36] M. Goldstein, J. Chem. Phys. **51** (1969), 3728.

[37] T.R. Kirkpatrick, D. Thirumalai and P.G. Wolynes, Phys. Rev. A **40** (1989), 1045.

[38] J.-P. Bouchaud and G. Biroli, J. Chem. Phys. **121** (2004), 7347.

[39] V. Lubchenko and P.G. Wolynes, Ann. Rev. Phys. Chem. **58** (2007), 235–266.

Jorge Kurchan
Ecole Supérieure de Physique
   et Chimie Industrielles de la Ville de Paris
10, rue Vauquelin
F-75231 Paris Cedex 05, France
e-mail: jorge@pmmh.espci.fr

Glasses and Grains, 25–39

Colloidal Glasses

David A. Weitz

Abstract. Glasses and granular materials share many features in common: Both can flow under some conditions but form disordered solids under other conditions. The similarity is captured within the jamming phase diagram, which considers how the solid-like state is fluidized with decreasing density, increasing shear stress, and increasing agitation, due to temperature in the case of molecular glasses and to shaking or some other form of agitation in the case of granular materials. Colloidal particles also undergo both jamming and glass transitions. They have the advantage that they are thermalized by temperature and that the particles themselves are large enough to be directly visualized. Thus, the study of the glass transition in colloids can provide an interesting comparison between molecular glasses and granular materials. This paper reviews the properties of colloidal suspensions near the colloidal glass transition, and explores both the glass-like properties and the jamming properties of these materials.

1. Introduction

At first glance, the sand on a beach and the glass in a window are vastly different. Sand sticks to your feet and makes the floor of your house dirty, much to the dismay of your mother or spouse. The glass in your window keeps the sand out of your house when the wind blows, while letting you look out and admire the view of the beach, should you be so lucky as to have such a beautiful view. However, despite these differences, the two materials share a surprising number of attributes. Of course, the sand is primarily silica, and this can be made into glass. However, the similarities go much deeper.

A glass is a material that has the same structure as a liquid. What distinguishes it from a liquid is its very slow relaxation time. In a liquid, thermal motion causes the molecules to constantly move, and these fluctuations lead to relaxation of the structure on short time scales. This structural relaxation is directly manifested in the viscosity of the fluid; for a fluid to flow, its structure must relax, and this relaxation is reflected directly in its viscosity. As a liquid approaches its glass transition, this structural relaxation becomes increasingly slow, ultimately taking longer than any experimental time scale. This leads to a dramatic increase in the viscosity of the liquid. Indeed, one definition of a glass is as a material where viscosity approaches 10^{12} Poise, which corresponds to flow that will take longer than a day [AMO86]. A glass can, in principle, be made to flow by applying a shear stress, although of course, many glasses will fracture or break before this happens. However, if the glass is not too brittle, and its elastic modulus is not too large, shear will lead to structural relaxation and the glass will flow. Alternatively, heating the glass above its glass transition will also lead to flow.

Sand, or any other granular material, is also a highly disordered structure [JNB96b, JNB96a]. However, it does not typically seem to be a fluid, so it is difficult to compare the structure of sand to that of a fluid. Nevertheless, if sand is shaken, it can flow like a fluid. This can be seen if a bucket of dry sand is turned gently on its side so the top of the sand begins to fall. If this is done slowly enough, the sand flows much like a fluid. An even better example of this is observed if air is gently blown up through sand [MD97]. If the flow rate is set exactly right, the level of the sand rises ever so slightly and the sand becomes fluidized, as is immediately seen if you try to run your hand through it; it feels much more like a fluid than a solid, and the grains of sand flow around your hand as you draw it through them. In this case, however, it is not thermal motion that is responsible for the fluidization, but instead it is the forces of the air blowing up through the grains. Thus, granular materials differ significantly from glasses because thermal energy is not sufficient to cause the individual grains to move; structural relaxation, and hence fluidization, can only occur through the addition of some larger form of energy, such as shaking or blowing air through the medium. However, if the grains are fluidized in this fashion, they can then explore their phase space, and they can behave very much like a fluid. Moreover, as this structural relaxation ceases, a granular material loses its ability to structurally relax, and retains the same structure it had, very much like a glass.

The similarity between glasses and granular material has been elegantly described through the concept of "jamming", and the associated jamming phase diagram [LN98]. The original jamming phase diagram has three control parameters, each of whose values defines an axis on the graph. One axis is the temperature, which provides a measure of how the system explores its phase space. For a glass, this is just the temperature of the system, whereas for a granular material some other mechanism must cause the system to explore its phase space and this effectively becomes the temperature. The second axis is a measure of the density. For a glass, this is a measure of the expansion of the system, whereas for a granular material, this is a measure of the volume fraction of the grains. The third axis is the amount of shear stress applied to the system. The temperature along an axis is plotted as the inverse of the system temperature, so a decrease in the system temperature leads to an increase along the axis. Similarly, the volume fraction or density is also plotted as the inverse, again leading to a lower volume fraction resulting in an increase along the axis of the system. When plotted in this fashion, the innermost quadrant becomes the jammed, or solid phase, and increasing along any three of the axes leads to a point where the system is fluidized. Thus there is a surface on the phase diagram that corresponds to the jamming transition for granular materials or the glass transition for glasses.

The jamming picture provides an elegant means to compare glasses and granular materials, and highlights the similarity in their behavior. However, several essential differences remain. Perhaps the most stark difference is the means by which each system is fluidized. A glass depends on temperature for its fluidization, and it falls out of equilibrium, and becomes a glass, as the temperature is lowered. By contrast, thermal energy has no effect on a granular material, as it is an athermal system; however, it too must be fluidized, and this must come from the addition of energy in some other form.

There is one system that shares many of the features of both glasses and granular matter. This system is a colloidal dispersion. Colloids are solid particles immersed in a fluid. The particles are small enough that they undergo Brownian motion. Typically the particles are less than a few microns in size. Their Brownian motion ensures that the colloidal particles sample their phase space and are therefore thermalized. However, the size of the particles is larger than that of the molecules in a typical glass, and the granularity of the particles has important implications for their behavior. Thus, colloidal particles fall between traditional molecular glasses and traditional granular materials. They have the behavior of "granular glasses". Their behavior offers insight into both traditional glasses and granular materials. This paper will explore some examples of this.

Colloidal particles are of great interest in themselves, as they can model the behavior of many complex fluid systems. However, even more interesting is the scale of the particles and the control that is possible over their properties. The particles can be synthesized with exquisite precision, leading to nearly perfect monodispersity in particle size, with variations of less than a few percent in their radii. Moreover the interparticle interactions can be precisely controlled and easily tuned. Perhaps the most intriguing feature of their study is that their size, being around one micron, is ideal for visualization in an optical microscope [VBW95]; moreover, the time scale of their motion is sufficiently slow that their dynamics can be followed in real time. A typical time scale is given by the time it takes to move their own diameter; since the particles are in a fluid, their microscopic motion is diffusive, and a typical diffusion time, given by $\tau = a^2/D_0$, ranges from msec to sec. Here, a is the particle radius and $D_0 = k_B T/6\pi\eta a$ is the Stokes-Einstein diffusion coefficient of an isolated particle in solution, where k_B is Boltzmann's constant and η is the viscosity of the fluid. With modern confocal microscopy, it is now possible to follow the motion of individual particles over time in 3D; up to 10,000 particles can be followed simultaneously. Because the particles are all the same shape and are spherical, their diffraction pattern is known; thus the precision with which the location of the particle centers can be located is given by signal-to-noise, rather than the more commonly assumed Rayleigh diffraction limit; thus it is possible to identify the location of each particle to within a few percent of its radius [CG96]. This offers the opportunity to explore their dynamics with unprecedented precision, and provides new insight into the behavior of these "granular glasses."

2. Repulsive colloidal glasses

In order to ensure that colloidal particles remain suspended, it is essential that they do not stick together upon random collisions that invariably occur. Thus, all stable colloidal colloids possess some form of repulsive interaction between the particles. The simplest of these is through volume exclusion; the particles can not occupy the same volume. This requires a very strong, yet very short-range repulsive interaction between particles. There is no interaction except when two particles touch, and then the interaction is very strongly repulsive. Such colloidal particles then behave as hard spheres [PVM86].

Hard-sphere colloidal particles exhibit a glass transition [PVM87]. However, since there is no repulsive energy between the particles, except when they exactly touch, the enthalpic term of the free energy can be neglected, and the only contribution to the free energy of the system is entropic. Moreover, the temperature of the suspension can only be changed a relatively small amount before its properties are significantly modified. For example, the continuous fluid can either boil or freeze, and in either case, the colloidal suspension would no longer behave as a suspension. In fact, much smaller temperature changes usually suffice to drastically modify the properties of the suspension, and as a result, temperature is typically not a good control parameter for colloidal particles. Instead, the control parameter is the entropy, and this is controlled by varying the particle volume fraction, ϕ. Perfectly uniform, or monodisperse, spherical particles will undergo a crystallization transition at $\phi = 0.494$, and will coexist with a fluid-like order of the particles up to $\phi = 0.55$, whereupon the sample will remain crystalline up to the maximum packing of the face-centered cubic lattice that forms, $\phi = 0.74$ [PVM86]. However, if the sample is rapidly quenched to a higher volume fraction, the structural relaxation that is essential for it to undergo crystallization is suppressed: The particles become increasingly crowded as ϕ increases, making it increasingly difficult for the particles to move and to undergo structural relaxation. As the particles become increasingly crowded, the structural relaxation time becomes increasingly larger. This behavior has all the hallmarks of a glass transition [PVM87]. This has been rather widely investigated as the behavior is well described using mode-coupling theory, which describes the shape of the relaxation as measured with light scattering, and correctly accounts for the divergence of the structural relaxation time [VMU93].

One of the most direct manifestations of the onset of a glassy state is shown by the mechanical response of the suspension, which can be measured with a rheometer [MW95]. We use a sample of silica spheres suspended in ethylene glycol, where they interacted as hard spheres. The particle radius is $a = 0.21 \mu$m, with a polydispersity in radius of about 20%, which prevents crystallization. The sample is held between the walls of a double-walled Couette geometry, enabling sensitive measurement of the mechanical response. The rheometer is controlled-strain, and applies a known strain at a given frequency and measured the resultant stress at the same frequency. Frequency dependent measurements of the real, or elastic, $G'(\omega)$, and the imaginary, or viscous, $G''(\omega)$, moduli are shown respectively in the upper and lower parts of Figure 1. At the lower ϕ, $G''(\omega)$ is dominant, and both moduli increase with frequency. However, as ϕ increases, $G'(\omega)$ begins to dominate over an extended range of frequencies; moreover, it develops a plateau where it varies only very slowly with frequency, while $G''(\omega)$ exhibits a definite and reproducible minimum. At higher frequencies, both moduli begin to increase, with $G''(\omega)$ rising more sharply, ultimately overtaking $G'(\omega)$. We can describe this behavior using the same theoretical approach that was used to account for the structural relaxation. We use a mode-coupling theory that also includes a contribution of the fluid at high frequencies, where it must begin to dominate the measured response. The results are in excellent agreement with the data, as shown by the solid lines in Figure 1 [MW95]. This is direct evidence that the colloidal suspension behaves like a glass.

The important feature to note in the data in Figure 1 is the fact that the elastic, or storage modulus, $G''(\omega)$, is greater than the viscous, or loss modulus, $G''(\omega)$, over an extended range of frequencies, including those that are most "apparent" to common use. This implies that the material is behaving like a solid, and does exhibit a shear modulus, over these frequencies. At lower frequencies, the material undergoes structural relaxation; this relaxation extends to lower and lower frequencies as the volume fraction of particles increases, and will ultimately become so large that it can not be measured with these techniques: This is the glass transition. These data highlight the fact that the control parameter for colloidal hard spheres is volume fraction rather than temperature, and that an increase in ϕ corresponds to a decrease in T, reflecting the correspondence of ϕ^{-1} to T. Because there is no interparticle interaction energy, the solid-like behavior of the particles must result from purely entropic origins.

To explore the origin of this elasticity, we use confocal microscopy to visualize the particles as the glass transition is approached [WCL+00]. This allows us to monitor the motion of the particles, and to observe the nature of the structural relaxation. For these experiments, we use poly-(methylmethacrylate) particles, sterically stabilized by a thin layer of poly-12-hydroxystearic acid [AGH86]. The particles have a radius $a \approx 1.18\mu m$, a polydispersity of $\approx 5\%$, and are dyed with rhodamine and suspended in a cycloheptylbromide/decalin mixture which nearly matches both the density and the index of refraction of the particles. We track particles for the entire duration of the experiment. We determine ϕ for each sample by measuring the volume per particle directly with the microscope.

The motion of the individual particles is very revealing: We plot the trajectories of several particles as they evolve in time in Figure 2; the grey shading indicates depth in the third dimension. Each particle is trapped in some local volume, with

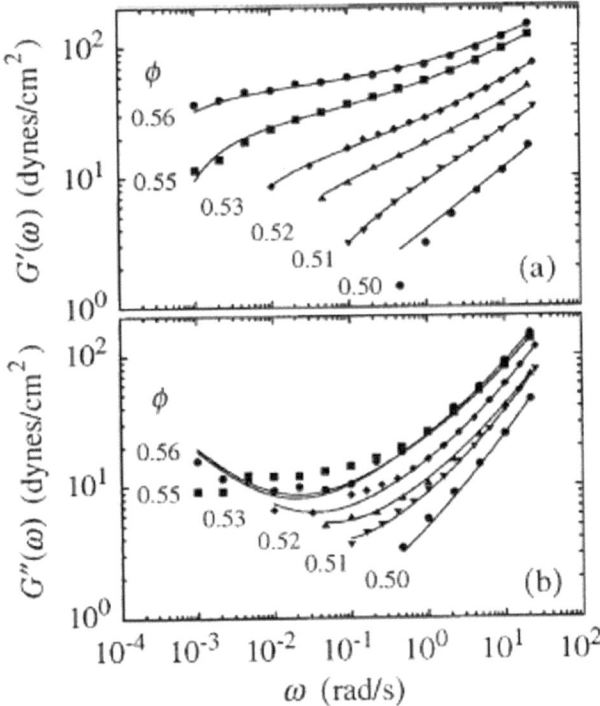

FIGURE 1. The frequency dependences of the (a) storage and (b) loss moduli for different volume fractions. All the measurements were performed at sufficiently low strains to be in the linear regime. The solid lines represent the fit to the mode-coupling model.

its trajectory moving only a small fraction of its size, until, at some random time, the particle moves a much larger amount, whereupon it is again trapped in a more localized volume. All particles that we investigate share this same characteristic motion. This motion reflects the nature of the structural relaxation. Physically, each particle is trapped in a cage composed of its neighbors, due to the crowding of the particles. Over the course of this cage-trapping time, the mean square displacement of the particle is almost independent of time; this also corresponds directly to the frequencies over which the elastic modulus exhibits the plateau in Figure 1. Eventually, the cage of neighbors surrounding the particle relaxes sufficiently that it moves, and when it does, it moves a larger distance. This results in structural relaxation of the system, and corresponds to the lowest frequencies in Figure 1, where the elasticity falls. These results provide important insight into the origin of the elasticity: During the time that the particle is trapped in its cage, its most probable position is within this cage; a small shear strain, as is applied to measure the elasticity, distorts the shape of the cage, reducing the number of

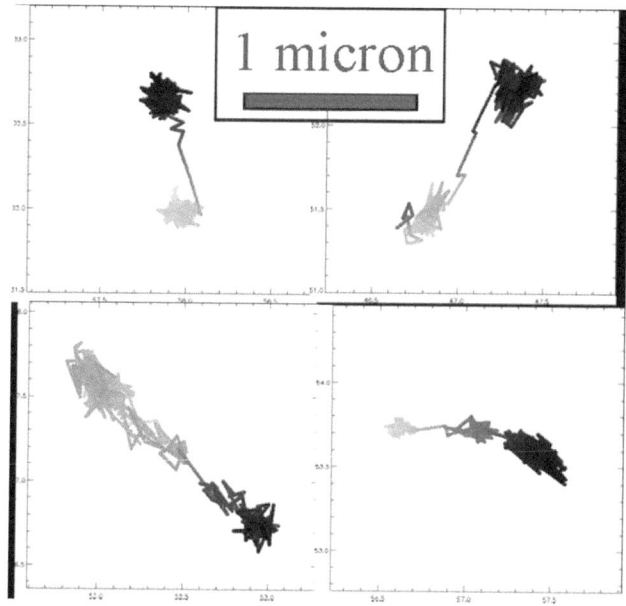

FIGURE 2. Temporal traces of the position of four different particles in a sample with a volume fraction of $\phi = 0.56$, near the glass transition, exhibiting cage trapping and cage escape. Grey shading indicated depth.

configurations that the particle can sample by straining the cage. This results in an increase in energy. If the strain is removed before the cage relaxes, the larger number of configurations are restored, and the energy is reduced. This storage of energy is directly reflected in the elasticity of the sample. It is strictly entropic in origin.

The characteristic time of the structural relaxation can be determined from the frequency of the relaxation measured with the rheological measurements, or from the mean square displacement of the particles, which exhibits an upturn at times corresponding to the structural relaxation frequency [WCL+00, WW02]. If we measure the motion of the particles on that time scale, we highlight the motion corresponding to the structural relaxation. When this is done, we observe that the structural relaxation of the particles is not uniformly distributed in space; instead, the motion of the relaxing particles is highly correlated among neighbors. To demonstrate this, we plot the particles in uniform time intervals of five minutes, and show those particles that have moved a large amount, and hence have undergone structural relaxation in the previous time interval, as red particles, drawn to scale, while those particles that have remained trapped are plotted as blue particles, and are drawn smaller, enabling the full 3D image to be visualized. Surprisingly, the moving particles are very strongly correlated in space, as shown in Figure 3. When one particle moves, it clearly opens a space behind

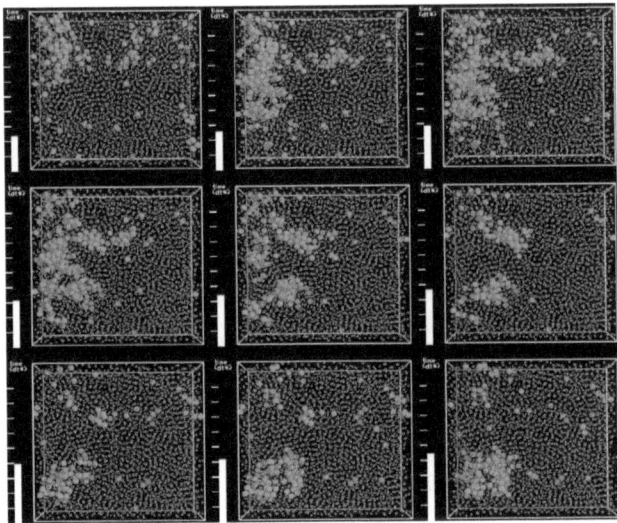

FIGURE 3. Correlation of particles undergoing structural relaxation in previous time step. The moving particles are in red and are drawn to scale, while the particles trapped in their cages, and hence more stationary in the previous time step, are shown in blue and are drawn smaller to allow the full 3D image to be visualized. The time separation between each image is five minutes.

it, through which a neighboring particle can move. This large scale correlation between the relaxing particles is what drives the longer relaxation times; it is not sufficient to have only a single particle move, but rather a large number of particles must move collectively to enable structural relaxation. We can define a length scale for these correlated motions by measuring the average size of regions of nearest-neighbor particles that undergo this structural relaxation. The average size increases dramatically as the volume fraction approaches the colloidal glass transition, $\phi_g \approx 0.58$, but then decreases to essentially zero above ϕ_g, as show in Figure 4. The structural relaxation is clearly very heterogenous in both space and time. Similar behavior occurs in molecular glass formers, where this motion is referred to as dynamic heterogeneities [SRS91, CE95]. However, direct observation of such behavior is not feasible with molecular glass formers, whereas with colloidal particles it is readily seen. It is also seen in computer simulation [DGP99].

When the sample is well within the glassy region, the structural relaxation occurs much more rarely, and when it does occur, it tends to be more highly localized. The characteristic time scale between relaxation events gets very long, corresponding to a large energy barrier for such a relaxation event to occur. Indeed, it is virtually impossible to observe these events in any reasonable experimental time frame. However, they can be sped up through application of a shear

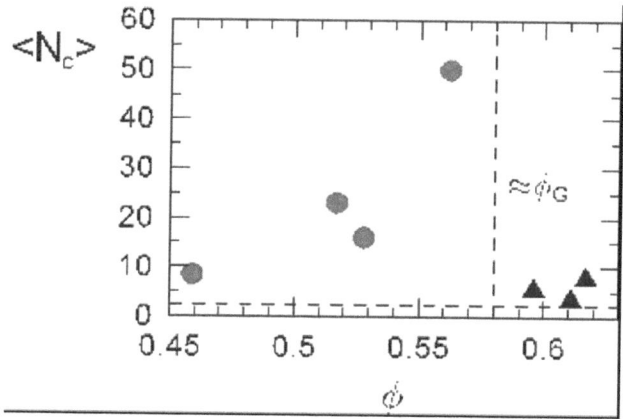

FIGURE 4. Average size of the correlated regions of particles undergoing structural relaxation. The size increases sharply as ϕ_g is approached, and then falls precipitously above.

strain. If this is done sufficiently slowly, local structural relaxation events are observed [FL98, SWS07].

Similar structural relaxation must also occur in granular materials, although these are more difficult to study as they are more difficult to observe in 3D. In addition, for granular materials, the dynamic heterogeneity is not thermally activated as it is in a glassy material that has not quite reached the glass transition. By contrast, similar more localized structural relaxation must occur in granular material that is flowing due to a shear stress, such as grains of sand flowing down an incline. These effects are probably best studied using colloidal particles, or these "granular glasses."

3. Attractive colloidal glasses

The interparticle interaction between the colloidal particles can be very precisely controlled. Stability of the colloidal particles against aggregation demands a very strongly repulsive interaction between the particles. For hard-sphere colloidal particles this is achieved by a very strong short range repulsion. For these repulsive particles, the phase behavior is directly determined by the volume occupied by the particles and by their crowding as ϕ increases and approaches ϕ_g. Such crowding only occurs when the volume fraction of particles fills nearly all space. However, colloidal particles are unique in that they can also have a weak attractive interaction between particles. This attraction must be sufficiently weak to ensure that the particles do not permanently stick to one another and aggregate. The attraction is typically at energy scales of only a few $k_B T$. However, this attractive interaction means that the system can become solid-like with a more spatially heterogeneous structure. This is commonly called gelation, but this shares many of the features of a glass transition. This behavior is actually quite common in technological uses of colloidal suspensions or other complex fluids, and thus the study of the properties of colloidal gelation is of importance both for practical purposes as well as an example of a glass-like transition.

This sort of colloidal glass transition, or gelation, is typically a very strongly kinetic process. The particles are destabilized and begin to aggregate; if the attraction is very large, there is no possibility for the particles to come apart once they have stuck to one another, and the aggregation is irreversible. In this case, the motion of the particles prior to sticking to one another has very important consequences for the resultant nature of the solid structure or gel. In the limit of diffusive particle motion and diffusion-limited collisions, the aggregates formed are fractal, with a fractal dimension of $d_f = 1.8$; as a result, the aggregates become more and more tenuous, and, on average, have a lower and lower density, as they grow [WO84]. Thus, the sample can form a solid gel [CG92, BMGW92] at arbitrarily low volume fractions, and the actual lower limit of ϕ that will gel is set by other factors, namely by the intrinsic strength of the gel which withstands the thermal motion, which becomes larger with larger length scales [MCT+04]. This process is known as diffusion-limited gelation. By contrast, if the attractive interaction is not as strong, there is some possibility that the bonds between particles can break, and as a result, the solid-like transition occurs at higher concentration, whose value depends on the magnitude of the attractive energy. In this case, the particles can be very susceptible to sedimentation due to buoyancy mismatch; as a result, there are many reports in the literature of behavior which is clearly affected by gravitational collapse [PPIB94]. This tends to obscure the underlying behavior, and thus experiments with buoyancy-matched samples are essential to fully understand the properties.

The existence of a well-defined boundary between fluid-like and solid-like states is most clearly shown by the rheological properties of the samples. The rheological behavior of weakly attractive particles exhibits a remarkable property: The data for every sample can be scaled onto a single master curve [TW00], as shown for samples of carbon black in oil at different volume fractions in Figure 5. The interaction energy between particles can also be varied through addition of different concentrations of a dispersant, a surfactant-like molecule that adsorbs on the surface of the particles and imparts some steric stability. This results in scaling of the rheological response onto exactly the same master curve. For a fixed interaction energy and changing volume fraction, those samples with a higher ϕ have a larger elastic modulus, and their data occupy the lower frequency side of the master curve. As ϕ decreases, the elastic modulus decreases, and the data fall more and more to the higher frequency side of the master curve. The actual origin of the scaling behavior can be understood by considering a model that simply adds two contributions to the response: a frequency-independent, elastic contribution that depends on the volume fraction of particles, and a viscous response that increases linearly with frequency that reflects the contribution of the background

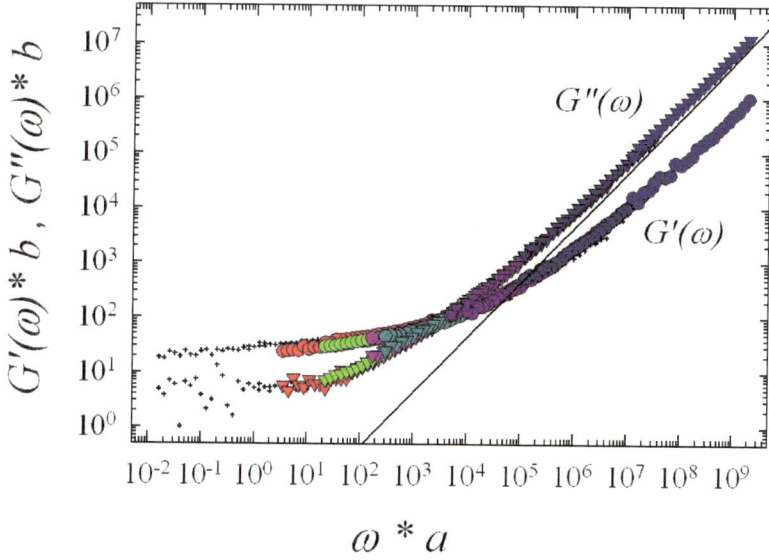

FIGURE 5. Scaling of the rheological response of samples of carbon black of different volume fractions. Similar behavior is observed for samples of a fixed volume fraction and varying interaction energy.

fluid to the response. Thus, the dominant response at low frequencies is the elastic response, whereas at high frequencies it is the viscous response. As a result, the data must be scaled along both axes, and the relationship of the two resultant scaling parameters should be linear; this is observed. Thus, this simple model describes the behavior quite well. The important consequence of this scaling is that even when the response is too weak to clearly measure the elastic component, as occurs, for example, for samples at very low volume fractions, it is nevertheless feasible to determine the elasticity by the scaling of the data. This allows us to quite precisely identify the boundary between a solid-like gel state and a fluid-like state.

Interestingly, the behavior of the weakly attractive colloidal particles exhibits exactly the same sort of behavior as that predicted for a "jamming" transition for granular particles [TPC+01]. Indeed, the original speculation about jamming also included a speculation that a jammed state should also exist for attractive particles; in this case, the attractive energy is what holds the system together, rather than the excluded volume of repulsive particles. Here, the control parameters are now $k_B T/U$, where U is a measure of the attractive energy of the interparticle potential, $1/\phi$ and σ, the shear stress on the sample which can cause it to fluidize. The measured jamming phase diagram for weakly attractive colloidal particles exhibits a hyperbolic shape, shown in Figure 6, rather than the concave shape originally predicted. Indeed, more recent work on jamming now predicts a shape closer to that observed for colloidal particles.

D.A. Weitz

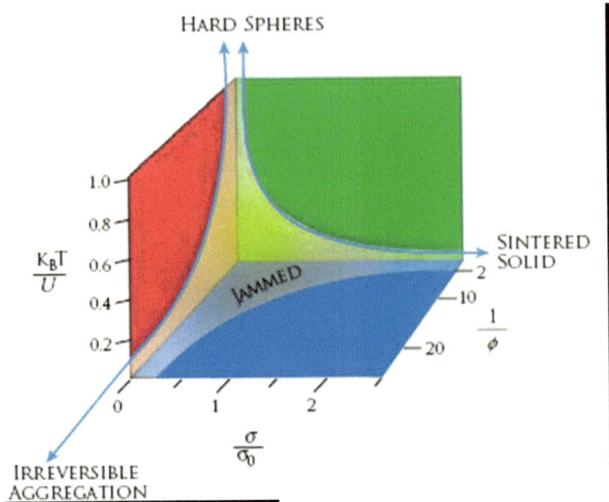

FIGURE 6. Jamming phase diagram for attractive colloidal particles.

Again, visualization of the individual colloidal particles provides important new insight into the behavior [LZC+08]. When the attractive interaction is very short range, the magnitude of the attractive interaction, as specified by k_BT/U, can be determined precisely by measuring the cluster mass distribution, measured for interaction energies below the gelation limit, where the clusters remain transient. These can be compared to predictions based either on analytic results obtained using a model interaction, or through simulations. Interestingly, the boundary of the solid-like behavior is always found to be exactly at the boundary of spinodal decomposition. This suggests that the gelation, which is strictly a kinetic phenomenon, is always preceded by spinodal decomposition, which is a quasiequilibrium phenomenon. Thus, even when the system is quenched well into the gel state, the system must first pass through spinodal decomposition. This leads to phase separation into colloid-rich and colloid-poor regions, driven by spinodal decomposition. The colloid-rich region is always at high volume fractions, comparable to that of the attractive colloidal glass. As a result, the colloid-rich region undergoes a kinetic arrest, freezing in the gel state, as shown in Figure 7. Similar behavior is observed for all attractive colloidal systems [VAL+97]. Thus, this provides new insights into the behavior of colloidal gelation.

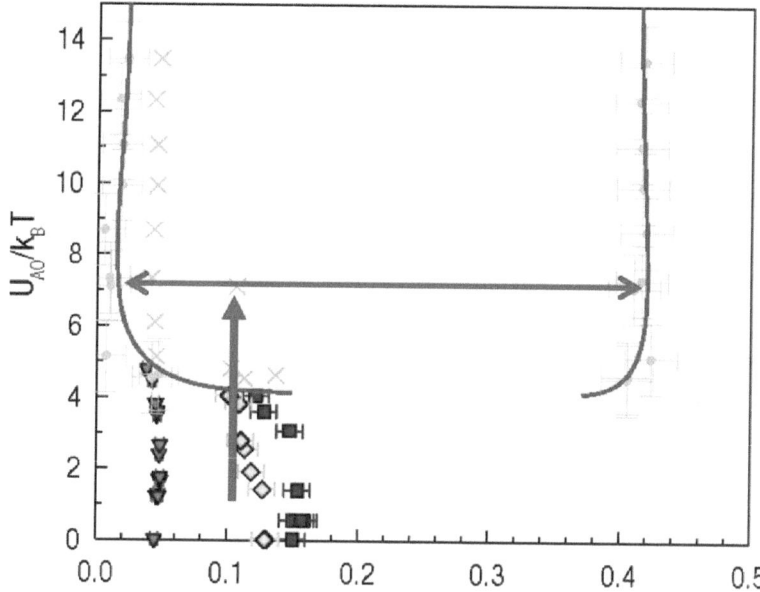

FIGURE 7. Model for gelation of attractive colloidal particles. A quench into the gel state initially causes spinodal decomposition into colloid-rich and colloid-poor regions. The colloid-rich region is always at the volume fraction of an attractive colloidal glass.

4. Conclusions

This paper has presented a very brief summary of results which describe the properties of colloidal particles viewed as "granular glasses," systems that share features of traditional molecular glass formers and granular materials. The wealth of behavior of such granular glasses is immense, and there are many more possible analogies that have been explored but not reported here. There are even more that are awaiting further work.

Acknowledgment

The work reported in this paper was carried out with many members of a large research team. Important contributions were made by John Crocker, Peter Lu, Tom Mason, Peter Schall, Francesco Scortino, Veronique Trappe, Eric Weeks and Emmanuella Zaccarrelli. The author is indebted to them all. The research was supported in part by NASA (NAG3-2284), the NSF (DMR-0602684) and the Harvard MRSEC (DMR-0820484). The author gratefully acknowledges this support.

References

[AGH86]   L. Antl, J.W. Goodwin, and R.D. Hill, *The preparation of polymethyl methacrylate latices in non-aqueous media*, Colloids Surf. **17** (1986), no. 1, 67–78.

[AMO86]   C.A. Angell, D.R. MacFarlane, and M. Oguni, *The Kauzmann paradox, metastable liquids, and ideal glasses*, Ann. NY Acad. Sci **484** (1986), 241–247.

[BMGW92] J. Bibette, T.G. Mason, H. Gang, and D.A. Weitz, *Kinetically induced ordering in gelation of emulsions*, Physical Review Letters **69** (1992), no. 6, 981–984.

[CE95]   M.T. Cicerone and M.D. Ediger, *Relaxation of spatially heterogeneous dynamic domains in supercooled ortho-terphenyl*, The Journal of Chemical Physics **103** (1995), 5684.

[CG92]   M. Carpineti and M. Giglio, *Spinodal-type dynamics in fractal aggregation of colloidal clusters*, Physical Review Letters **68** (1992), no. 22, 3327–3330.

[CG96]   J.C. Crocker and D.G. Grier, *Methods of digital video microscopy for colloidal studies*, Journal of Colloid and Interface Science **179** (1996), no. 1, 298–310.

[DGP99]   C. Donati, S.C. Glotzer, and P.H. Poole, *Growing spatial correlations of particle displacements in a simulated liquid on cooling toward the glass transition*, Physical Review Letters **82** (1999), no. 25, 5064–5067.

[FL98]   M.L. Falk and J.S. Langer, *Dynamics of viscoplastic deformation in amorphous solids*, Physical Review E **57** (1998), no. 6, 7192–7205.

[JNB96a]   H.M. Jaeger, S.R. Nagel, and R.P. Behringer, *Granular solids, liquids, and gases*, Reviews of Modern Physics **68** (1996), no. 4, 1259–1273.

[JNB96b]   _____, *The physics of granular materials*, Physics Today **49** (1996), no. 4, 32–39.

[LN98]   A.J. Liu and S.R. Nagel, *Jamming is not just cool any more*, Nature **396** (1998), no. 6706, 21–22.

[LZC⁺08]   P.J. Lu, E. Zaccarelli, F. Ciulla, A.B. Schofield, F. Sciortino, and D.A. Weitz, *Gelation of particles with short-range attraction*, Nature **453** (2008), no. 7194, 499–503.

[MCT⁺04]   S. Manley, L. Cipelletti, V. Trappe, A.E. Bailey, R.J. Christianson, U. Gasser, V. Prasad, P.N. Segre, M.P. Doherty, S. Sankaran, et al., *Limits to gelation in colloidal aggregation*, Physical Review Letters **93** (2004), no. 10, 108302.

[MD97]   N. Menon and D.J. Durian, *Particle motions in a gas-fluidized bed of sand*, Physical Review Letters **79** (1997), no. 18, 3407–3410.

[MW95]   T.G. Mason and D.A. Weitz, *Linear viscoelasticity of colloidal hard sphere suspensions near the glass transition*, Physical Review Letters **75** (1995), no. 14, 2770–2773.

[PPIB94]   P.N. Pusey, W.C.K. Poon, S.M. Ilett, and P. Bartlett, *Phase behaviour and structure of colloidal suspensions*, Journal of Physics Condensed Matter **6** (1994), 29–29.

[PVM86]   P.N. Pusey and W. Van Megen, *Phase behavior of concentrated suspensions of nearly hard colloidal spheres*, Nature **320** (1986), no. 6060, 340–342.

[PVM87]   _____ , *Observation of a glass transition in suspensions of spherical colloidal particles*, Physical Review Letters **59** (1987), no. 18, 2083–2086.

[SRS91]   K. Schmidt-Rohr and H.W. Spiess, *Nature of nonexponential loss of correlation above the glass transition investigated by multidimensional NMR*, Physical Review Letters **66** (1991), no. 23, 3020–3023.

[SWS07]   P. Schall, D.A. Weitz, and F. Spaepen, *Structural rearrangements that govern flow in colloidal glasses*, Science **318** (2007), no. 5858, 1895.

[TPC⁺01]   V. Trappe, V. Prasad, L. Cipelletti, P.N. Segre, and D.A. Weitz, *Jamming phase diagram for attractive particles*, Nature **411** (2001), no. 6839, 772–775.

[TW00]   V. Trappe and D.A. Weitz, *Scaling of the viscoelasticity of weakly attractive particles*, Physical Review Letters **85** (2000), no. 2, 449–452.

[VAL⁺97]   N.A.M. Verhaegh, D. Asnaghi, H.N.W. Lekkerkerker, M. Giglio, and L. Cipelletti, *Transient gelation by spinodal decomposition in colloid-polymer mixtures*, Physica A **242** (1997), 104–118.

[VBW95]   A. Van Blaaderen and P. Wiltzius, *Real-space structure of colloidal hard-sphere glasses*, Science **270** (1995), 1177–1177.

[VMU93]   W. Van Megen and S.M. Underwood, *Glass transition in colloidal hard spheres: Mode-coupling theory analysis*, Physical Review Letters **70** (1993), no. 18, 2766–2769.

[WCL⁺00]   E.R. Weeks, J.C. Crocker, A.C. Levitt, A. Schofield, and D.A. Weitz, *Three-dimensional direct imaging of structural relaxation near the colloidal glass transition*, Science **287** (2000), no. 5453, 627.

[WO84]   D.A. Weitz and M. Oliveria, *Fractal structures formed by kinetic aggregation of aqueous gold colloids*, Physical Review Letters **52** (1984), no. 16, 1433–1436.

[WW02]   E.R. Weeks and D.A. Weitz, *Properties of cage rearrangements observed near the colloidal glass transition*, Physical Review Letters **89** (2002), no. 9, 95704.

David A. Weitz
Department of Physics & SEAS
Harvard University
29 Oxford St.
Cambridge, MA 02138, USA
e-mail: `weitz@seas.harvard.edu`

Glasses and Grains, 41–76
© 2011 Springer Basel AG

Glass and Jamming Transitions

Giulio Biroli

Abstract. This is an introductory chapter to glass and jamming transitions. We present basic facts as well as recent discoveries, such as dynamical heterogeneities. In the final part we discuss a prominent theoretical approach: the Random First-Order Transition Theory. In order to convey the state of the art and the running debates, we make use both of the usual presentation style of physics papers but also of a dialogue format.

1. Introduction

Glasses are one of the materials most known and used by humans. Obsidian – a volcanic glass – was used for prehistoric tools and weapons. Now we easily design glasses with desired mechanical or optical properties on an industrial scale; glasses are widely present in our daily life and even used to create art objects, such as Murano glasses. Yet, a microscopic understanding of the glassy state of matter remains a challenge for statistical physicists. Glasses share similarities with crystalline solids, they are both mechanically rigid, but also with liquids since they both have similar disordered structures at the molecular level. In 1995 P.W. Anderson wrote that, *the deepest and most interesting unsolved problem in solid state theory is probably the nature of glass and the glass transition.* The aim of this article is to give an introduction to this problem, describe the mysteries related to it, show the reasons why it is so much studied by theoretical physicists and, finally, explain a theory that can possibly lead to its solution.

Since this is still an unsolved problem there are several – at least apparent – contradictions in the literature, and several debated questions and biases in interpreting the experimental results, and even in deciding which ones are important and which ones are not. In order to convey the state of the art and some of the debated questions, I will make use both of the usual presentation style of physics papers but also of a dialogue format between two characters: Salviati, a theoretical physicist, and Cleverus, a laymen – but a very clever one![1]

---

[1] As the reader certainly knows, such dialogues are not new in the physics literature. Even some characters, like Salviati, have appeared already several times [1, 2]. Of course, I have no ambition

2. Basic facts

The usual way to obtain a glass consists in cooling a liquid. The quench must be fast enough in order to avoid the standard first order phase transition towards the crystalline phase. The metastable phase reached in this way is called 'supercooled phase'. In this regime the viscosity and the relaxation timescale, e.g., the decorrelation time of the density field, increase by more than 14 orders of magnitude: from picoseconds to hours. This is a quite unusually broad range for a condensed matter system. Depending on the cooling rate, typically of the order of 0.1–100 K/min, and the patience of the people carrying out the experiment, typically hours, the supercooled liquid falls out of equilibrium at a certain temperature, called the glass transition temperature T_g, and becomes an amorphous rigid material – a glass.

This experimental glass 'transition' is clearly *not* a thermodynamic transition at all, since T_g is only empirically defined as the temperature below which the material has become too viscous to flow on a 'reasonable' timescale. Pragmatically, physicists have agreed upon a common definition of T_g as the temperature at which the shear viscosity is equal to 10^{13} Poise (also 10^{12} Pa.s). In order to grasp how viscous this is, recall that the typical viscosity of water at ambient temperature is of the order of 10^{-2} Pa.s. How long would one have to wait to drink a glass of water with a viscosity 10^{14} times larger?

The increase of the relaxation timescale of supercooled liquids is remarkable not only because of the large number of decades involved but also because of its temperature dependence. This is vividly demonstrated by plotting the logarithm of the viscosity (or the relaxation time) as a function of T_g/T, as in Figure 1 [3]. This is called the 'Angell' plot and is very helpful in classifying supercooled liquids. A liquid is called strong or fragile depending on how the viscosity changes as a function of T_g/T in the Angell plot. Straight lines correspond to 'strong' glass-formers and to an Arrhenius behaviour. In this case, one can extract from the plot an effective activation energy, suggesting quite a simple mechanism for relaxation by 'breaking' locally a chemical bond. The typical relaxation time is then dominated by the energy barrier to activate this process and, hence, has an Arrhenius behaviour. Window glasses fall in this category. The terminology 'strong' and 'fragile' is not related to the mechanical properties of the glass but to the evolution of the short-range order close to T_g. Strong liquids, such as SiO_2, have a locally tetrahedric structure which persists both below and above the glass transition contrary to fragile liquids whose short-range amorphous structure disappears rapidly upon heating above T_g. If one tries to define an effective activation energy for fragile glass formers using the slope of the curve in Figure 1, then one finds that this energy scale increases when the temperature decreases, a 'super-Arrhenius' behaviour. This increase of energy barriers immediately suggests that the glass formation is a collective phenomenon for fragile supercooled liquids: many degrees of freedom have to move cooperatively to make the system relax and this

whatsoever to share anything with the great scientists who wrote those dialogues, except maybe the fun in writing them.

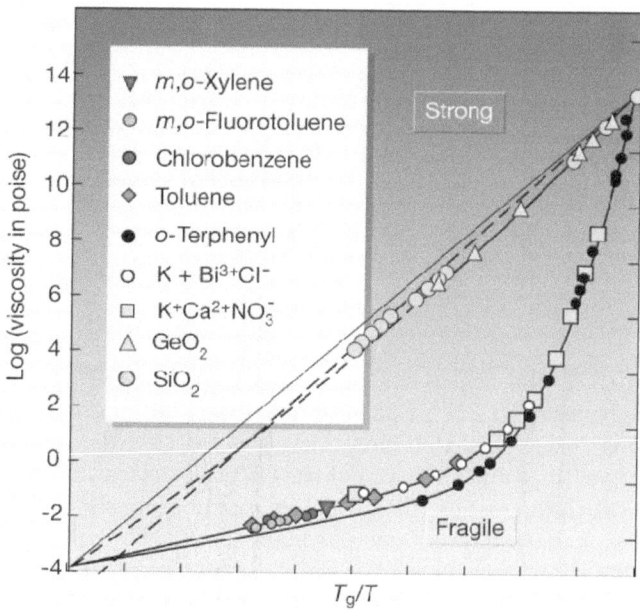

FIGURE 1. Angell plot of the viscosity of several glass-forming liquids approaching the glass temperature T_g [3]. For 'strong' liquids, the viscosity increases in an Arrhenius manner as temperature is decreased, $\log \eta \sim E/(K_B T)$, where E is an activation energy and the plot is a straight line, as for silica (SiO_2). For 'fragile' liquids, the plot is bent and the effective activation energy increases when T is decreased towards T_g, as for ortho-terphenyl. Note that there exists a continuous range of liquids from very fragile to very strong. By plotting their corresponding data one would fill the gap between the two curves in the figure above.

leads to growing energy barriers. Support for this interpretation is provided by the fact that a good fit of the relaxation time or the viscosity is given by the Vogel-Fulcher-Tamman law (VFT):

$$\tau_\alpha = \tau_0 \exp \left[\frac{DT_0}{(T - T_0)} \right], \tag{1}$$

which suggests a divergence of the relaxation time and of the effective barrier $DT_0 T/(T - T_0)$ and, hence, a phase transition of some kind at a temperature T_0. A smaller D in the VFT law corresponds to a more fragile glass. Note that there are other comparably good fits of these curves, such as the Bässler law,

$$\tau_\alpha = \tau_0 \exp \left(K \left(\frac{T_*}{T} \right)^2 \right), \tag{2}$$

Substance	o-terphenyl	2-methyltetra-hydrofuran	n-propanol	3-bromopentane
T_g	246	91	97	108
T_0	202.4	69.6	70.2	82.9
T_K	204.2	69.3	72.2	82.5
T_K/T_0	1.009	0.996	1.028	0.995

TABLE 1. Values of glass transition temperature, VFT singularity and Kauzmann temperatures T_K for four supercooled liquids [4].

that only leads to a divergence at zero temperature but still implies a divergent effective energy barrier, $K(T^*)^2/T$, and hence some kind of growing cooperativity. Actually, although the relaxation time increases by 14 orders of magnitude, the increase of its logarithm, and therefore of the effective activation energy, is modest, and experimental data do not allow one to unambiguously determine the true underlying functional law beyond any reasonable doubt. For this and other reasons, physical interpretations in terms of a finite temperature phase transition must always be taken with a grain of salt.

However, there are other experimental facts that shed some light and reinforce the use of Eq. (1). Among them, there is an empirical connection found between kinetic and thermodynamic behaviours. Consider the part of the entropy of the liquids, S_{exc}, which is in excess compared to the entropy of the corresponding crystal, and plot it as a function of T. As for the relaxation time, one cannot follow this curve below T_g in thermal equilibrium. However, extrapolating the curve below T_g apparently indicates that the excess entropy vanishes at some finite temperature, called T_K[2]. The big surprise is that T_K is generically very close to T_0, the temperature at which a VFT fit diverges. This coincidence is quite remarkable: for materials with glass transition temperatures that vary from 50 K to 1000 K, the ratio T_K/T_0 remains close to 1, up to a few percents[3]. Examples reported in Ref. [4] are provided in Table 1. This link between kinetics and thermodynamics appears also in the empirical relation found by Adam-Gibbs, which connects the relaxation time to the excess entropy mentioned above, and reads: $\ln \tau(T) \propto [TS_{xs}(T)]^{-1}$. The Adam-Gibbs relation holds reasonably well both for single liquids as temperature is varied, and cross-sectionally for different liquids. Were this relation exact, it would indeed imply that the temperature at which the relaxation time diverges must coincide with the one at which the excess entropy vanishes.

But where do these empirical relations, in particular the one between T_0 and T_K, come from? This is a first mystery to elucidate. Note that the 'coincidence'

---

[2]The sub-index K stands for Kauzmann, who first discussed the possible existence of T_K.

[3]Note, however, there are some liquids where T_K and T_0 differ by as much as 20%, and so a perfect correlation between the two temperatures is not established experimentally [5].

between T_0 and T_K strongly suggests that some kind of phase transition may well take place at $T_0 = T_K$. This underlying "ideal glass transition"[4] would be the phenomenon responsible for the physical behavior of supercooled liquids and for the glass transition seen in experiments. But what kind of transition is this? And why is the excess entropy an important quantity anyway? Goldstein introduced a physical picture that has been indeed very instrumental in our understanding of the glass problem. He assumed – and this was later verified numerically – that approaching T_g the system of N particles (or molecules) forming the super-cooled liquid explores a part of the energy landscape (or configuration space) which is full of minima separated by barriers that increase[5] when temperature decreases. In the Goldstein scenario the dynamic evolution in the energy landscape would then consist in a rather short equilibration inside the minima, followed by 'jumps' between different minima. At T_g the barriers have become so large that the system remains trapped in one minimum, identified as one of the possible microscopic amorphous configurations of a glass. Following this interpretation, one can split the entropy into two parts. A first contribution is due to the fast relaxation inside one minimum, a second counts the number of metastable states, $S_c = \log N_{\text{metastable}}$, which is called the 'configurational' entropy. Assuming that the contribution to the entropy due to the 'vibrations' (or coarse-grained vibrations) around an amorphous glass configuration is not very different from the entropy of the crystal[6], one finds that $S_{\text{exc}} = S_{\text{liquid}} - S_{\text{crystal}} \approx S_c$. Within this approximation, T_K corresponds to a temperature at which the configurational entropy vanishes. Concomitantly, the properties of the energy landscape visited by the system change drastically, since below T_K the system becomes stuck in a handful of low-lying states. Assuming that S_c vanishes linearly as suggested by extrapolations of S_{exc} and using the thermodynamic relation $T \frac{dS}{dT} = C_p$, one finds a downward jump of the specific heat C_p at T_K and therefore a truly thermodynamic phase transition.

At this point the reader might have reached the conclusion that the glass transition may not be such a difficult problem: there are experimental indications of a diverging timescale and a concomitant singularity in the thermodynamics. It simply remains to find static correlation functions displaying a diverging correlation length related to the emergence of 'amorphous order'. This would allow us to classify the glass transition as a standard second-order phase transition. Remarkably, this remains an open and debated question despite several decades of research. Simple static correlation functions are quite featureless in the supercooled regime, notwithstanding the dramatic changes in the dynamics. A simple

---

[4]It is called ideal because, if it exists, it would be a true phase transition, compared to the cross-over taking place at T_g, and it could be reached only if one were able to equilibrate the systems on arbitrarily large, meaning in practice unattainable, timescales.

[5]Note that the energy landscape does not change with temperature. The regions of the landscape sampled by the system, instead, do depend on temperature. At low temperature, the sampled regions are characterized by higher barriers in Goldstein's scenario.

[6]Note, however, that the above assumptions should not be taken for granted, see for instance the recent discussions in [6, 7, 8].

static quantity is the structure factor defined by

$$S(q) = \left\langle \frac{1}{N} \delta \rho_{\mathbf{q}} \delta \rho_{-\mathbf{q}} \right\rangle, \tag{3}$$

where the Fourier component of the density reads

$$\delta \rho_{\mathbf{q}} = \sum_{i=1}^{N} e^{i\mathbf{q} \cdot \mathbf{r}_i} - \frac{N}{V} \delta_{\mathbf{q},0}, \tag{4}$$

where N is the number of particles, V the volume, and $\mathbf{r}_i$ is the position of particle i. The structure factor measures the spatial correlations of particle positions, but it does not show any diverging peak in contrast to what happens, for example, at the liquid-gas tri-critical point where there is a divergence at small $\mathbf{q}$. A snapshot of a supercooled liquid configuration in fact just looks like a glass configuration, despite their widely different dynamic properties. What happens then at the glass transition? A more refined understanding can be gained by studying dynamic correlations or response functions.

A dynamic observable studied in light and neutron scattering experiments is the intermediate scattering function,

$$F(\mathbf{q}, t) = \left\langle \frac{1}{N} \delta \rho_{\mathbf{q}}(t) \delta \rho_{-\mathbf{q}}(0) \right\rangle. \tag{5}$$

Different $F(\mathbf{q}, t)$ measured by neutron scattering in supercooled glycerol [9] are shown in Figure 2 for different temperatures. These curves show a first, rather fast, relaxation to a plateau followed by a second, much slower, relaxation. The plateau is due to the fraction of density fluctuations that are frozen on intermediate timescales, but eventually relax during the second relaxation. The latter is called 'alpha-relaxation', and corresponds to the structural relaxation of the liquid. This plateau is akin to the Edwards-Anderson order parameter, q_{EA}, defined for spin glasses, which measures the fraction of frozen spin fluctuations. Note that q_{EA} continuously increases from zero below the spin glass transition. Instead, for structural glasses, a finite plateau appears above any transition.

Many other remarkable phenomena take place when a supercooled liquid approaches the glass transition. I presented above some of the most important ones, but many others have been left out for lack of space.

To sum up, approaching the glass transition, a liquid remains stuck for a very long time in an amorphous configuration. It eventually evolves but this takes such a long time that below a certain temperature this regime cannot be probed anymore. As a matter of fact, the number of microscopic configurations in which the glass can get stuck is exponentially large as shown by the value of the configurational entropy at T_g: a few k_B per particle. There seems to be a very large degeneracy in the ways molecules can arrange themselves such as to form mechanically stable, amorphous patterns around which they vibrate without exploring all the other stable patterns. The way in which this degeneracy decreases with temperature seems to be tightly connected to the way the relaxation times increase. Possibly, an ideal

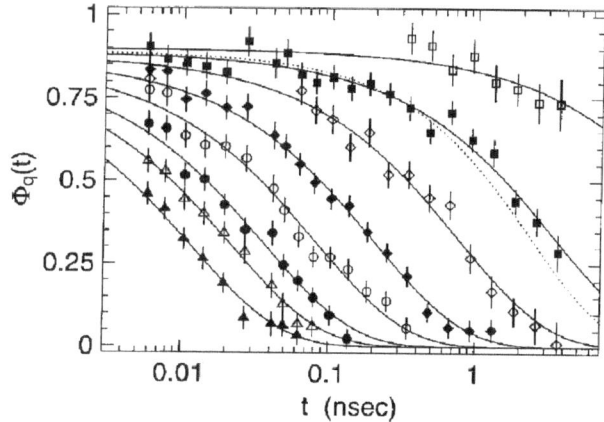

FIGURE 2. Temperature evolution of the intermediate scattering function normalized by its value at time equal to zero for supercooled glycerol [9]. Temperatures decrease from 413 K to 270 K from right to left. The solid lines are fit with a stretched exponential with exponent $\beta = 0.7$. The dotted line represents another fit with $\beta = 0.82$.

glass transition takes place at $T_0 = T_K$. At this temperature the configurational entropy, a measure of this degeneracy, would vanish and the relaxation timescale would diverge. This exponential degeneracy of the "phases" towards which the liquid can freeze is a feature that classical theories of phase transitions cannot easily handle, and that requires new tools. At the same time, this feature seems to be the very essence of glassiness: in order to prevent fast crystallisation, the interaction between molecules must be able to generate enough "frustration" to make the energy landscape rough and rocky and trap the system in a configuration not very different from an arbitrary initial configuration of the liquid.

3. The jamming-glass transition of colloids and grains

Colloidal suspensions consist of big particles suspended in a solvent. The typical radii of the particles are in the range $R = 1$–500 nm. The solvent, which is at equilibrium at temperature T, renders the short-time dynamics of the particles Brownian. The microscopic timescale for this diffusion is given by $\tau = R^2/D$ where D is the short-time self-diffusion coefficient. Typical values are of the order of $\tau \sim 1$ ms, and thus are much larger than the ones for molecular liquids (in the picosecond regime). The interaction potential between particles depends on the system, and this large tunability makes colloids very attractive objects for technical applications. A particularly relevant case, on which we will focus in the following, is a purely hard sphere potential, which is zero when particles do not overlap and infinite otherwise. In this case the temperature becomes irrelevant, apart from a trivial rescaling of the microscopic timescale. Colloidal hard spheres systems have been intensively studied in experiments, simulations and theory varying their density ρ, or their volume fraction $\phi = \frac{4}{3}\pi R^3 \rho$. Hard spheres display a fluid phase from 0 to intermediate volume fractions, a freezing-crystallisation transition at $\phi \simeq 0.494$, and a melting transition at $\phi \simeq 0.545$. Above this latter value the system can be compressed until the close packing point $\phi \simeq 0.74$, which corresponds to the FCC crystal. Interestingly, a small amount of polydispersity (particles with slightly different sizes) suppresses crystallization. In this case, the system can be more easily 'supercompressed' above the freezing transition without nucleating the crystal, at least on experimental timescales. In this regime the relaxation timescale increases very fast. At a packing fraction $\phi_g \simeq 0.58$ it becomes so large compared to typical experimental timescales that the system does not relax anymore: it is jammed. This 'jamming transition' is obviously reminiscent of the glass transition of molecular systems and indeed several studies have shown that the phenomenon that take place increasing the volume fraction are analogous to the ones seen in molecular supercooled liquid: the relaxation timescales increase very fast and can be fitted by a VFT law in density as in Eq. (1), dynamical correlation functions display a broad spectrum of timescales and develop a plateau as in Figure 2, no static growing correlation length has been found, etc. Also the phenomenon of dynamic heterogeneity that we will address later is present in both cases.

It is important to stress that the super-compressed state of colloids is really akin to the super-cooled state of molecular liquids: it is metastable with respect to the crystal but otherwise at equilibrium. We will not address here problems in which the colloid ages or is compressed extremely fast.

In theoretical studies of the jamming-glass transition it is customary to neglect the hydrodynamics effects and focus on hard spheres mixtures undergoing a Brownian dynamics. It is a well-known result that the probability distribution at long times for such a system converges to the equilibrium Boltzmann distribution.

As a summary, comparing the jamming-glass transition of colloids to the glass transition of liquids, one finds several astonishing similarities, despite the fact that the systems are rather different in nature: for the former the dynamics is Brownian and the order parameter of the Jamming-Glass transition is the density, whereas for the latter the dynamics is Newtonian and the order parameter is the temperature. Another major important difference is that experiments in colloids can only track the first five decades of slowing down because the microscopic timescale for colloids is very large. An important consequence is that the comparison between glass and colloidal transitions must be performed by focusing in both cases on the first five decades of the slowing down. This would correspond to relatively high temperatures in molecular liquids. Understanding how much and to what extent the glassiness of colloidal suspensions is related to the one of molecular liquids is an active domain of research.

Another class of systems that display a Jamming-Glass transition which share very similar properties with the glass transition of molecular liquids are driven

FIGURE 3. A bi-dimensional, bi-disperse granular material, composed of about 8.000 metallic cylinders of diameter 5 and 6 mm in equal proportions, is sheared quasi-statically in a horizontal deformable parallelogram. The shear is periodic, with an amplitude $\theta_{max} = \pm 5°$. The volume fraction ($\phi = 0.84$) is maintained constant by imposing the height of the parallelogram. The dynamics of 2818 grains located in the center of the device is followed by a High Resolution Digital Camera which takes a picture each time the system is back to its initial position $\theta = 0$. The microscopic unit of time is one cycle, a whole experiment lasting 10.000 cycles. Experiments with this setting have been performed and analyzed in publications by Dauchot et al., see this volume, page 137.

granular media. If colloids can be thought of as siblings of molecular liquids, grains must be thought as some more distant relatives, let's say cousins. The reason is that grains are macroscopic objects and, as a consequence, do not have any thermal motion. A granular material is therefore frozen in a given configuration if no energy is injected into the system. However, it can be forced in a steady state by an external drive, such as shearing or tapping. An example of an experimental setting studied in the group of O. Dauchot at CEA Saclay is shown in Figure 3. It consists in a bi-dimensional and bi-disperse granular material, which is quasi-statically sheared in a horizontal deformable parallelogram. The dynamics in the steady state is quite different in nature from both the equilibrium dynamics of colloids and molecular liquids: energy is continuously injected into the system and subsequently dissipated. Time-averaged observables cannot be obtained from an equilibrium Boltzmann measure. Actually, the steady state probability distribution is generally unknown. Despite these facts, steady state dynamics of granular systems at high density show remarkable similarities with the ones of colloids and molecular liquids. First, the timescales for relaxation of the density field and for diffusion of a tagged particle increase very fast when density is increased, without any noticeable change in structural properties. It is now established that many phenomenological properties of the glass transition also occur in granular assemblies. Going beyond the mere analogy and understanding how much these different physical systems are related is a very active domain of research. Actually, since the underlying dynamics and microscopic properties are so different between liquids and grains, it would be highly non-trivial to find that the microscopic mechanism responsible for the corresponding glass transitions are the same or even very similar. We will discuss some very recent and astonishing results at the end of this paper.

4. First dialogue

Here we present a first dialogue about the glass transition between two characters: Salviati, a theoretical physicist, and Cleverus, a very clever laymen. Cleverus is just back from a cruising trip around the globe with his sail boat, while Salviati has been actively working on the glass transition lately[7].

Cleverus: Hi Salviati, long time no see. How're you doing? I'm just back from a long sailing trip. As you know I like to bring some scientific stuff to read during these trips. This time I brought with me several papers and books on the glass transition. It looks like a hot topic. Recently, there was even an article about it in the New York Times! Well, finally, I didn't have much time. I read some stuff though. I didn't understand much and I have a lot of questions. First, why you guys are so much interested in this problem? Liquids stop to flow at low enough temperature and become glasses. So what? It looks like a very boring problem to me. Why is this phenomenon difficult to explain?

Salviati: Hi Cleverus, I see you are fine and provocative as usual. You asked many questions already, let me start from the last one. In order to answer it, let me quote the Nobel laureate P.W. Anderson: "We are so accustomed to the rigidity of solid bodies – the idea, for instance, that when we move one end of a ruler the other end moves the same distance – that it is hard to realize that such an action at a distance is not built into the laws of nature except in the case of the long-range forces such as gravity and electrostatics." We understand why crystals are rigid. This is related to the spontaneous symmetry breaking of translation invariance: particles arrange on a periodic lattice and is the lattice that transmits the force. Of course this reasoning works for crystal, but why liquids become amorphous *rigid* materials at the glass transition? Is there a related spontaneous symmetry breaking transition? As Anderson put it: "We are so accustomed to this rigidity property that we don't accept its almost miraculous nature, that is an 'emergent

---

[7]Of course, Salviati does not represent the author. Resemblances to well-known scientists are coincidental, except for the ones you find out.

property' not contained in the simple law of physics, although it is a consequence of them". Why glasses are rigid is a very difficult question.

Cleverus: I see, when you are in trouble you start quoting Anderson. I still don't get it. If I consider the strong liquids you were describing before, then it looks clear to me why they become rigid: each particle has to jump a barrier in order to move. When the temperature becomes much less than the barrier the time for this process is of the order of the age of the universe, so all particles are stuck in their positions and the material is a solid. Concerning rigidity, well, the barriers for motion are there, because particles are arranged on an amorphous lattice and the interaction with their almost fixed neighbours creates a barrier for motion for a given particle. So, the reason of rigidity is the same as for crystals.

Salviati: Yes, you might be right for strong liquids. But there are still several mysteries related to the process of glass-formation. First, even if your explanation might work for strong glasses, it does not for fragile ones. In those systems, the barrier *is increasing* with temperature. Why is it so? Furthermore, you seem to accept disingenuously that the low temperature phase of the liquid is a broken symmetry state, where particles are arranged in an amorphous way. But this is something completely new in physics! As for the emergence of chaos from purely deterministic dynamics, here we have low disordered energy states that emerge cooling a completely homogeneous high temperature liquid. Imagine, that a thermodynamic transition takes place, what kind of transition would it be? Moreover, the number of amorphous configurations in which the glass can be stuck is huge. That is something unheard of in usual theories of phase transitions.

Cleverus: OK, right, I start to see why the problem is difficult and even why *you* are interested. Maybe it's a new kind of phase transition that you cannot handle and understand with the usual theory of phase transitions. But, still, why should it be interesting for a laymen like me. I tried to understand it from the New York Times article. I read a lot of amusing quotes but still didn't find a reason to be interested.

Salviati: Maybe, another reason to find it interesting is that glasses are the archetype of complex systems. Their understanding, or more generally the research on glassy systems, could lead to several spin-offs in other fields like computer science, finance, biology.

Cleverus: Mhmm, don't look at me with your wide blue eyes as the snake in the Walt Disney's "Jungle Book". Maybe doing that and saying "Bzzzzz Glasses Bzzzzz Complex Bzzzzzz Interesting" works with your friends. It does not with me, pal. So you've got to explain me a little better. By the way I thought you guys used the words "complex systems" just in your grant proposals. Do you really mean it when you say that glasses are complex systems? What does it mean "complex system" anyway?

Salviati: I won't try to give a general definition of what is a complex system. That looks like a difficult thing to do. I would say, as the Supreme Court justice Potter Stewart that didn't know how to define pornography: I don't know how to define it but I know it when I see it.

Cleverus: Uhh, is that a joke? I would have indeed trouble defining "complex systems" but pornography ...

Salviati: Cleverus, com'on, stop teasing me. I would say that glassy liquids are complex because the properties of their low energy states are extremely difficult to obtain, in particular by a dumb and local optimization procedure. Imagine that you want to find the most compact structure for hard spheres. Yes, don't look at me that way, I know that is just a crystal. But forget it for a moment and think what a local optimization procedure would do or, even better, what a particle subject to thermal noise would do. The particle would try to pack close to itself as many neighboring spheres as possible. And its fellow spheres would do the same. Well, doing this, one does not end up with a crystal on large scale in three dimension[8] but in a relatively loose packing. Imagine now that you want to obtain very compact amorphous packings, which is the counterpart of a "low energy state" for hard spheres. Then you have to optimize the way in which particles are packed together on a large scale. And in this case you don't know how to do it a priori contrary to the crystal case. If you wait that the system does it by itself just by thermal fluctuations, then you have to wait an extremely long time since many particles have to rearrange cooperatively.

This kind of difficulty in finding low energy states is common to many other problems in science where many degrees of freedom, agents, boolean variables interact in a contradictory way (we use the term frustration to denote this phenomenon). In those cases the macroscopic properties that emerge are very much unexpected and very often impossible to predict from heuristic or simple arguments. And, by the way, I wasn't trying to selling you an old car when I talked about spin-offs in other branches of science. For example, a central problem in computer science, random K-SAT, was recently analyzed in full detail with techniques developed for glassy systems. The authors even showed that random K-SAT is characterized by a glass transition [19]!

Cleverus: OK, I understand a little bit better your point now. I would have many questions on complex systems in finance and biology but let's stick to glasses. I don't understand why you made all that fuss about the behavior of fragile liquids. Maybe the explanation is just the same one I gave you for strong glasses, except that the *local* barrier increases approaching T_g.

Salviati: As I told you, there appears to be evidence that the barrier is actually diverging. Our experience taught us that this may happen only when there is some kind of collective phenomenon. A local barrier cannot diverge. You should read the paper by Montanari and Semerjian [18], they proved that a diverging timescales at finite temperature is necessarily related to a diverging length.

Cleverus: Actually, it was among the papers I brought with me. But, oh boy, it's difficult for me. I tried hard for two days. One night, while I was sailing in the arctic ocean, I was so focused on it, or maybe I fell asleep I don't remember, that I almost crushed my boat against an iceberg. Ending my trip in a Titanic

---

[8]It does in two dimension because there is no *frustration*.

way didn't sound like a good idea, so I decided to wait and ask you for a full explanation. Please, go to the blackboard and explain.

Salviati: Well, ehm, it would take some time. I will give you now an imprecise poor man's explanation – *un argument à deux balles* – as the french say. I strongly suggest you to read their paper afterwards. Imagine that you have a system characterized by a diverging relaxation time. If there is no growing cooperation between particles, or no growing correlation lengths whatsoever, then the system can be thought of as formed by independent pieces. Each small microscopic sub-system would then relax independently and have the same diverging relaxation time. On the other hand a finite system, as a rule of thumb, cannot have a relaxation time larger than e^{aN} where N is the number of its degrees of freedom and a is a constant. Thus, if you start from the hypothesis that the relaxation time is diverging at finite temperature then you are bound to conclude that there must be some kind of cooperativity such that the macroscopic system cannot really be considered the union of many microscopic independent parts.

Cleverus: Alas, my dear Salviati, you don't know whether there is an ideal glass transition and the time is really diverging. You just see it increasing. So, although I like your explanation, you cannot really use it to prove that the glass transition is not a just a complete local phenomenon. So, all your mumbo jumbo about a new kind of phase transition, collective phenomena, etc. could well be plain wrong.

Salviati: OK, you are right to be skeptical. But now we know for sure that the glass transition is not just a completely local phenomenon. In the last decades the research has focused on what we call the real space properties of the dynamics. We have found that dynamical processes leading to relaxation are correlated on length-scales that increase approaching the glass transition. Let me explain this in detail because it is a very important point.

5. Dynamical heterogeneities and dynamical correlations

A common feature of the way particles, grains or molecules move in glassy liquids is the so-called "cage effect". This is shown in Figure 4 and means that dynamical trajectories become very intermittent temporally: a particle typically rattles for a long time inside a "cage" formed by its neighbors and then it moves abruptly to a new position, around which it starts to rattle again. In order to show that the glass transition is not just a complete local phenomenon one has to study the way in which these "jumps out of the cage" (or cage jumps) are organized in space-time. In particular, are they correlated or completely independent, as would be the case for a completely local process? To answer this question, let us directly focus on their correlation or, more generally, on the correlation of local dynamical relaxations. In order to do that, let us write a global correlation $C(0,t)$ as a sum of local terms:

$$C(0,t) = \frac{1}{V} \int d^3r \, c(\mathbf{r}; 0, t), \qquad (6)$$

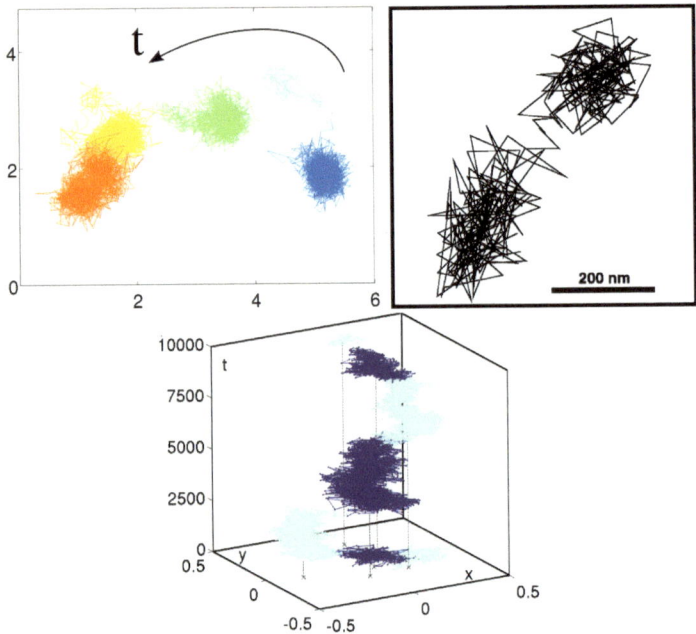

FIGURE 4. These are three examples of the cage effect in glassy dynamics. The three figures show typical particle trajectories approaching the glass and jamming transitions for a liquid (left: obtained by simulation [10]), a colloid (center: obtained by experiments [11]) and a granular system (right: obtained by experiments [12]). In all cases the size of the "cage" is some tenth of the average interparticle distance. Different colors are used to highlight different cages.

where V is the volume of the system. $C(0, t)$ could be the intermediate scattering function considered previously or could be a measure of mobility: $C(0, t) = \frac{1}{N} \sum_i Q_i(t) = \frac{1}{N} \sum_i e^{-[\mathbf{r}_i(t) - \mathbf{r}_i(0)]^2 / (2a^2)}$. In the latter case

$$c(\mathbf{r}; 0, t) = \frac{V}{N} \sum_i \delta(\mathbf{r}_i - \mathbf{r}) e^{-[\mathbf{r}_i(t) - \mathbf{r}_i(0)]^2 / (2a^2)},$$

where a is equal to the particle distance or, alternatively, some fraction of it.

The correlations in the dynamics can be probed by using the correlation function [13]:

$$G_4(\mathbf{r}; 0, t) = \langle c(\mathbf{0}; 0, t) c(\mathbf{r}; 0, t) \rangle - \langle c(\mathbf{0}; 0, t) \rangle \langle c(\mathbf{r}; 0, t) \rangle, \qquad (7)$$

Since $c(\mathbf{r}; 0, t)$ is generally a two-point function, G_4 is a four-point function. This is the reason for the sub-index 4. Direct numerical studies in glass-forming liquids and experimental studies in granular systems have shown that this function, evaluated

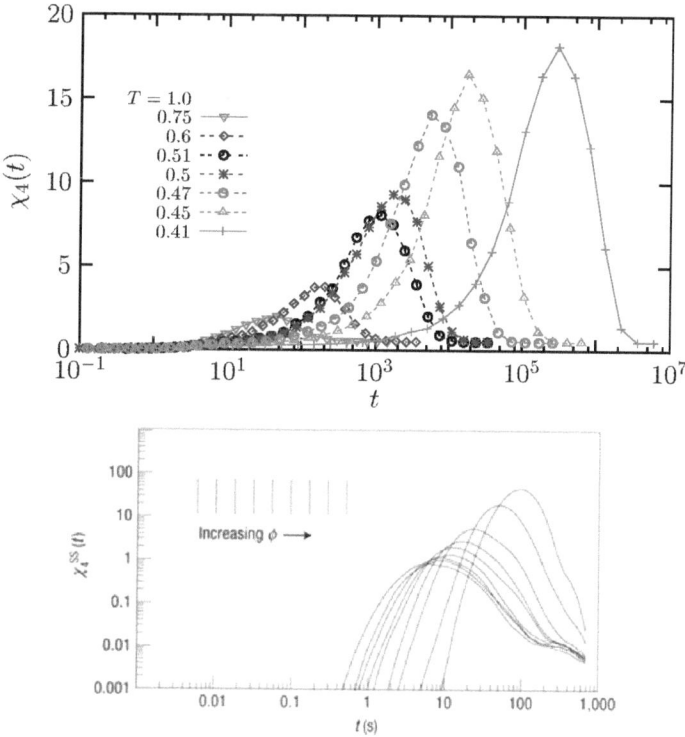

FIGURE 5. Time dependence of $\chi_4(t)$ quantifying the spontaneous fluctuations of the intermediate scattering function in a Lennard-Jones supercooled liquid [14] and in a granular system consisting in air-driven steel spheres [15]. The peak of $\chi_4(t)$ increases by lowering the temperature in the former case, and increasing the density in the latter one. This reveals that the glassy dynamics becomes more and more correlated approaching the glass and jamming transition.

at t equal to the relaxation time, becomes more and more long ranged approaching the glass or the jamming transition. However, these measurements are difficult and extracting precisely a length ξ_d from G_4 is complicated. The majority of works have instead focused on its integral, which is called $\chi_4(t)$. This is akin to a standard dynamical susceptibility in usual critical phenomena if one interprets $C(0,t)$ as the order parameter for the glass transition. It reads:

$$\chi_4(t) = \rho \int d^3 r G_4(\mathbf{r};0,t) = N\langle [C(t,0) - \langle C(t,0)\rangle]^2 \rangle. \qquad (8)$$

This function shows a very similar behavior in molecular liquids, hard spheres and grains. We show in Figure 5 the typical shape and the typical evolution of χ_4 for liquids and granular media (hard spheres behave similarly). As a function of time $\chi_4(t)$ first increases, it has a peak on a timescale that tracks the structural relaxation timescale and then it decreases. The peak value, which measures the volume on which the structural relaxation processes are correlated, is found to increase approaching the glass and jamming transitions. Note that if the dynamically correlated regions were compact, the peak of χ_4 would be just proportional to ξ_d^3 in three dimensions.

A visual characterization of dynamical correlations can be obtained by plotting a color map of $Q_i(\tau^*)$ where τ^* is the relaxation time of the system. In Figure 6 we plot this map for a two-dimensional glass-forming liquid (right) and a two-dimensional granular system close to jamming (left). There are clearly well-formed spatial structures. Their extension is what is probed by G_4 and χ_4. This phenomenon has been called dynamical heterogeneity. Note that independent local relaxations would instead lead to a homogeneously blurred figure displaying no spatial structure.

In molecular liquids, contrary to the other glassy systems, $\chi_4(t)$ has been measured only by numerical simulations and far from the glass transition because probing G_4 or χ_4 in experiments is beyond current capabilities. The reason is that one can track the particle positions for colloids and grains. This is instead impossible for molecular liquids. There are two ways to overcome this difficulty. First, it has been argued that non-linear responses should be directly related to χ_4 and these have been measured recently by the group of D. L'Hôte in CEA Saclay. Second, one can obtain by fluctuation-dissipation relations and rigorous inequalities a relationship between the way the relaxation timescale increases and the growing of dynamical correlations. The main idea is to obtain a rigorous lower bound on $\chi_4(t)$ using the Cauchy-Schwarz inequality $\langle \delta H(0) \delta C(0,t) \rangle^2 \leqslant \langle \delta H(0)^2 \rangle \langle \delta C(0,t)^2 \rangle$, where $H(t)$ denotes the enthalpy at time t. By using fluctuation-dissipation relations the previous inequality can be rewritten[9] as [16]

$$\chi_4(t) \geq \frac{T^2}{C_P} \left[\chi_T(t) \right]^2, \tag{9}$$

where the multi-point response function $\chi_T(t)$ is defined by

$$\chi_T(t) = \left. \frac{\partial F(t)}{\partial T} \right|_{N,P} = \frac{N}{T^2} \langle \delta H(0) \delta C(0,t) \rangle. \tag{10}$$

As a consequence, one can obtain a lower bound on the increase of dynamical correlations from the way the correlation function changes with temperature. This is useful for two reasons. First, it is easy to check[10] that it implies that a diverging timescale at a finite temperature must be accompanied by a diverging χ_4.

---

[9]Henceforth we will take $k_B = 1$.

[10]In order to do that, assume that the correlation function has a scaling behavior $f(t/\tau(T))$ and that $\tau(T)$ diverges at a finite temperature T_0. Plugging this expression into Eq. (9) one finds that the lower bound diverges and this implies a diverging χ_4 at T_0.

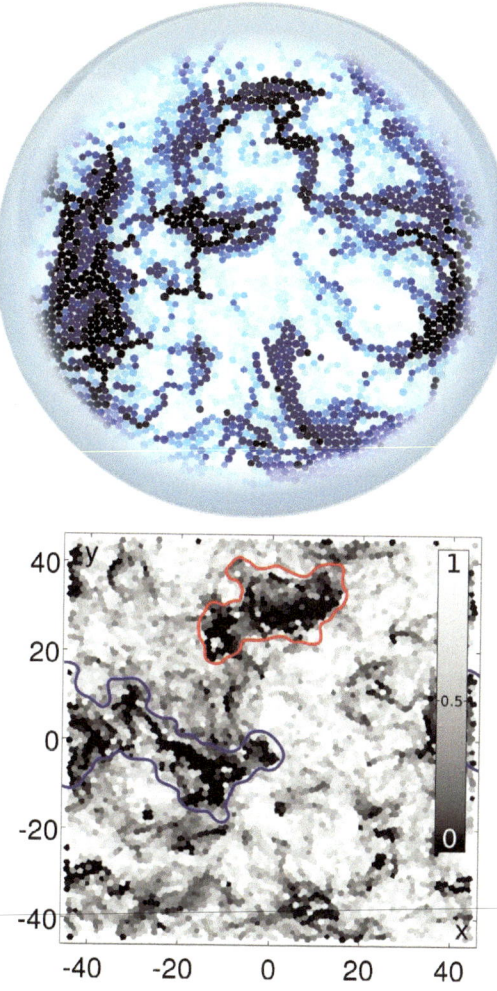

FIGURE 6. Color map of $Q_i(\tau^*)$ for: (Right) a $2D$ non-equimolar binary mixture of particles interacting via purely repulsive potentials of the form $u_{ab}(r) = \varepsilon(\sigma_{ab}/r)^{12}$ and (Left) the two-dimensional granular system described in Figure 3. Lowest values of $Q_i(\tau^*)$ correspond to particles that have already relaxed on the time τ^* and to darker colors.

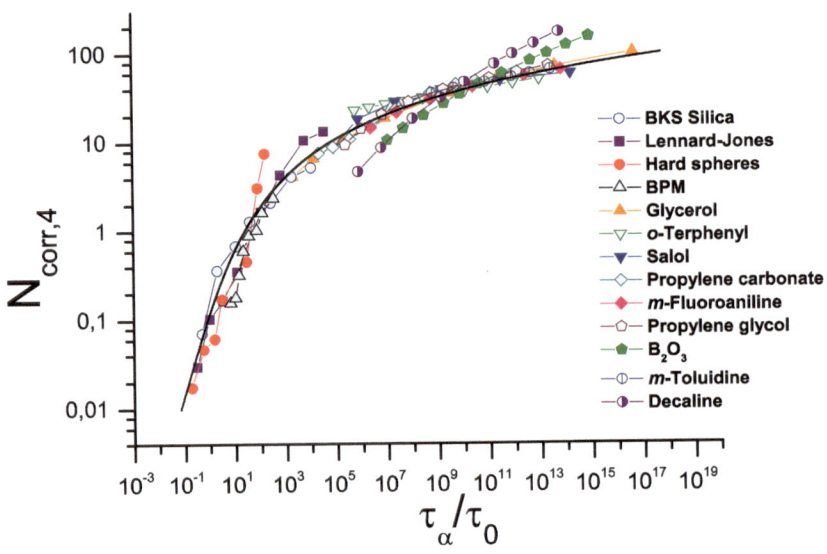

FIGURE 7. Universal dynamic scaling relation between number of dynamically correlated particles, $N_{\mathrm{corr},4}$, and relaxation timescale, τ_α, for a number of glass-formers [17], determined using Eq. (9).

Second, it actually provides a quantitative way to estimate, via a lower bound, dynamical correlations in glass-forming liquids close to the glass transition. Using this method, Dalle-Ferrier *et al.* [17] have been able to obtain the evolution of the peak value of χ_4 (or more precisely its estimation by the lower bound[11]) for many different glass-formers in the entire supercooled regime. In Figure 7 we show some of these results as a function of the relaxation timescale. The value on the y-axis, the peak of χ_4, is a proxy for the number of molecules, $N_{\mathrm{corr},4}$ that have to evolve in a correlated way in order to relax the structure of the liquid. Note that χ_4 is expected to be equal to $N_{\mathrm{corr},4}$, up to a proportionality constant which is not known from experiments, probably explaining why the high temperature values of $N_{\mathrm{corr},4}$ are smaller than 1. Figure 7 also indicates that $N_{\mathrm{corr},4}$ grows faster when τ_α is not very large, close to the onset of slow dynamics, and a power law relationship between $N_{\mathrm{corr},4}$ and τ_α is good in this regime ($\tau_\alpha/\tau_0 < 10^4$). The growth of $N_{\mathrm{corr},4}$ becomes much slower closer to T_g. A change of six decades in time corresponds to a mere increase of a factor about 4 of $N_{\mathrm{corr},4}$, suggesting logarithmic rather than power law growth of dynamic correlations.

---

[11]Numerical analysis suggested that the lower bound provides a rather good estimation of the rate of growth of χ_4 approaching the glass transition.

6. Second dialogue

Cleverus: Glassy dynamics have indeed a lot of interesting and remarkable real space properties. I agree now that any explanation of the glass transition in terms of a purely local process is bound to fail. Something I really find puzzling is the apparent similarity between liquids, colloids and grains. They are really different microscopically and yet ... You didn't seem to be very much surprised. Are all these glass-jamming transitions the same?

Salviati: You are right, I didn't stressed enough this similarity. Actually, I find it very remarkable too. As I said, one has to be careful not comparing apples to oranges: in colloids and grains one can probe only the first decades of slowing down. Thus, we don't really know whether their behavior is similar to the one of molecular liquids close to their glass transition, since there the relaxation time has grown of 14 or more decades. Still, at least in the first regime of slowing down of the dynamics, they really are similar and this is remarkable. I have to admit that we don't have a clear idea of why this is so. We have just indications by analytical [20] and numerical approaches [22] that indeed the glass transitions of hard spheres and molecular liquids may be driven by the same mechanism. We have to work more.

Cleverus: From your explanation on the real space properties of the dynamics I got that the dynamical slowing down is accompanied by a growing *dynamical* correlation length. This seems to suggest that the ideal glass transition – if such a thing exists – is purely dynamical; no thermodynamic transition, right?

Salviati: This is a tricky point [21]. Let me again quote Anderson and not because I feel in trouble! Already in 1983 he wrote: "This has, however, become a very knotty question. Some – but not all – transitions to rigid, glasslike states, may entail a hidden, microscopic order parameter which is not a microscopic variable in any usual sense, and describes the rigidity of the system. This is the fundamental difficulty of the order-parameter concept: at no point can one be totally certain that one can really exclude a priori the appearance of some new hidden order." He was right and well ahead of his time – as in many other cases. Only recently, we started to understand what this static correlation length could be. As a matter of fact, in the Montanari and Semerjian work we discussed before, they prove that the length which has to diverge together with the relaxation time is a *static one*!

Cleverus: Ahhh, I've got your point – I think. There could be a *static* correlation length, and the fact that you don't see it easily does not mean that it is not there. One has to be careful. But then, what is the relationship between the dynamic length you see, cf. your previous explanations, and the static one you don't?

Salviati: First, we have now direct evidences that there is a static growing length, I can tell you more later. As for your question on the relation between static and dynamic lengths, we didn't make up our minds. We tend to believe that if particles are correlated statically on a length ξ_s, then they will be dynamically correlated on a length that is at least equal, but possibly larger, than ξ_s. Actually, it is simple to construct models in which the dynamical length ξ_d is much larger than the static one. However, we don't have any definite proof that $\xi_d \geq \xi_s$, so it is probably better not to make any categorical or dogmatic statement.

Cleverus: I see. But do you guys have some hints of what is the theoretical explanation of the glass transition? I mean, if you don't even know whether it is static or purely dynamic . . .

Salviati: Well, don't be too negative. Of course, since the problem of the glass transition is not solved there are many different theoretical approaches, different schools of thought, . . .

Cleverus: Ah, there was a funny quote in the New York Times article. D. Weitz said: "There are more theories of the glass transition than there are theorists who propose them". Bravo, your experimental colleagues have a very good opinion of you.

Salviati: Mhmm, it is true that there are possibly 1001 derivations of the Vogel-Fulcher law. However there are very few real theories, meaning based on solid microscopic basis and able to explain a lot of properties of glassy liquids.

Cleverus: Why you cannot sort out which is the right one?

Salviati: The main difficulty is that all these theories are based on some kind of critical phenomenon. That's reasonable since there are indeed growing correlation lengths at the glass transition. The trouble is these lengths are not very large! You saw that the dynamical correlation length takes, at best, a value of 10 at the glass transition. The static lengths are probably smaller. This means that we are still rather far from the critical point, whatever it is. Hence, in order to obtain quantitative predictions that can be tested in experiments and to contrast theories, critical and universal results are not enough. One should be able to compute pre-asymptotic corrections. This is a fundamental difficulty of the glass transition: the relaxation timescale increases so fast that one cannot get close to the transition. This does not happen in standard critical phenomena because the timescale diverges as power law of $T - T_c$.

Cleverus: OK, well, you cannot compute pre-asymptotic effects. I hope you are at least able to compute the asymptotic ones.

Salviati: . . . it depends on the theory.

Cleverus: Oh com'on, stop to be so balanced and sphinxish in your answers. Tell me what theory you think is right and why.

Salviati: I have different reasons to believe that the most promising theory, at least the one which is the correct starting point, is the Random First-Order Transition Theory.

Cleverus: Ah, the RFOT theory! I tried to read many papers during my sailing trip. Oh boy, it was a nightmare. I had to choose between some that do awkward computations on very weird models: completely connected mean-field models with quenched disorder and multiple spins interactions. Why on earth these models should be related to glasses anyway? And others, with no computation at all. At first sight, I was happy with the latter, but then I found them full of so many, how do you call them – ah right – hand-waving arguments . . . One night after reading one of those papers, I dreamt to have the Hindu goddess Kali in front of me waving all her hands and explaining RFOT. It was a mesmerizing experience. But, it didn't help with RFOT. So, please, go ahead, explain it to me if you can.

7. The Random First-Order Transition Theory Part I: Mean-field theory

Historically, the origin of RFOT lies in the study of mean-field disordered systems and starts with pioneering works by Kirkpatrick, Thirumalai and Wolynes. Indeed the Random Energy Model of Derrida contains in a nutshell already many important aspects of RFOT. Many workers in the field of the glass transition often criticizes RFOT on the basis that is not clear why mean-field spin systems with quenched disorder have anything to do with glass-forming liquids, which are formed instead by interacting particles and devoid of quenched disorder. They are right. These systems are microscopically extremely different. However, they are possibly related in the same way liquids close to the liquid-gas transition are related to spin systems close to the ferromagnetic transition. They belong to the same universality class at least within mean-field theory. Nevertheless, it is clear that analyzing these spin models and, then, applying directly the results to real liquids can be baffling for researchers who want to understand how particles move and rearrange in real systems and are not used to mean-field disordered systems. At the very end, if RFOT is the theory of the glass transition, we must be able to explain it without making a long detour on quenched disordered systems. In the following, I shall endeavor to do that. As a consequence, alert and knowledgeable readers may find that I'm starting from the end instead of from the beginning and, at the same time, sweeping some difficult technical points under the rug. They are right.

7.1. A chaotic free energy landscape

We saw that close to T_g a liquid remains stuck in amorphous configurations and that the number of these is exponentially large in the system size. A mean-field theory of the problem should therefore be able to: (1) capture this kind of symmetry breaking (2) deal with an exponential number of states. As usual, the first approach to follow in order to get a handle on the physics, at least qualitatively, is mean-field theory. As for ferromagnets the first step is studying the evolution of the free-energy landscape. In the Curie-Weiss approach one computes the free-energy as a function of the global magnetization by a mean-field approximation. However, computing the free energy as a function of the global energy or density won't be enough for the glass transition. One has to be able to deal with the fact that the system can be stuck in many different amorphous configurations. As a consequence one is bound to compute the free-energy F as a function of the entire density field. F is defined using the Legendre transform. Consider for simplicity an interacting particle lattice model, the generalization to continuum systems is straightforward. In the lattice case a given configuration is determined by the number of particles, n_i, on every site i. In order to define F, one first introduces the thermodynamic 'potential'

$$W(\{\mu_i\}) = -\frac{1}{\beta} \log \sum_{\{n_i\}} \exp\left(-\beta H(\{n_i\}) + \sum_i \beta \mu_i n_i\right). \tag{11}$$

The free energy function $F(\{\rho_i\})$ is then defined as

$$F(\{\rho_i\}) = W(\{\mu_i^*\}) + \sum_i \mu_i^* \rho_i \tag{12}$$

where the μ_i^*s satisfy the equations $\frac{\partial W}{\partial \mu_i} + \rho_i = 0$ and, hence, are functions of all ρ_is. This is often called the TAP free energy because Thouless, Anderson and Palmer introduced it to study mean-field spin glasses. The generalization to continuum systems can be also performed by replacing the discrete variable n_i by a continuum density field $\rho(x)$. In this case F is called a density functional.

The free energy landscape is the hyper-surface generated by scanning F over all possible values of $\{\rho_i\}$. Its critical points, in particular the minima, play a crucial role. In fact by deriving the previous equation with respect to ρ_i one finds

$$\frac{\partial F}{\partial \rho_i} = \mu_i^*.$$

Thus, when there are no external fields (or local chemical potentials) the solutions of these equations are all the stationary points of the free energy landscape[12]. What are the main features of F for a system approaching the glass transition? This question of course cannot be answered exactly for a three-dimensional system. One has either to make use of approximations (as in the Curie-Weiss description of ferromagnets) or focus on mean-field lattices like Bethe lattices, which often provide a good approximation to the finite-dimensional ones. Furthermore, continuum systems are quite complicated; let us first focus on simple but reasonable lattice models.

A concrete example is given by 'lattice glass models'. These are models containing hard particles sitting on the sites of a lattice. The Hamiltonian is infinite if there is more than one particle on a site or if the number of occupied neighbors of an occupied site is larger than a parameter, m, and is zero otherwise. Tuning the parameter m, or changing the type of lattice, in particular its connectivity, yields different models. Lattice glasses are constructed as simple statmech models to study the glassiness of hard sphere systems and they have been shown by simulation to reproduce correctly the physics of glass-forming liquids (at least on the timescales accessible to simulations). The constraint on the number of occupied neighbors mimics the geometric frustration encountered when trying to pack hard spheres in three dimensions. Other models, which have a finite energy and, hence, are closer to molecular glass-formers, can be also constructed. The technical study

---

[12]For particle systems there is always a global chemical potential μ fixing the number of particles. In this case, one includes the global term $\mu \sum_i n_i$ in the definition of F so that all μ_i^* are zero.

of their free-energy landscape is quite involved. Here we will only discuss the main results and consider temperature as a control parameter. The generalization to models where the density or the chemical potential are the control parameters is straightforward.

The main result of the study of the free energy landscape is that it becomes 'rugged' at low temperature and characterized by many minima and saddle points. Actually, the number of minima is so large that in order to count them one has to introduce an entropy, called configurational entropy or complexity, $s_c = \frac{1}{N} \log \mathcal{N}(f)$, where $\mathcal{N}(f)$ is the number of free-energy minima with a given free energy density f. The density profile corresponding to one given minimum is amorphous and lacks any type of periodic long-range order, and different minima are ... very different. This is a first result, which is very welcome. Glasses freeze in an exponential number of different amorphous configurations, and within the mean-field approximation of simple but reasonable lattice models we indeed find a lot of amorphous free energy minima at low temperature. Assuming that all the minima are mutually accessible, one can compute the thermodynamic properties, i.e., the partition function, by summing over all states with their Boltzmann weights. Formally, one can introduce a free-energy dependent complexity, $s_c(f, T)$, that counts the number of TAP minima with free-energy density f at temperature T. The partition function of the system then reads:

$$Z(T) = \int \mathrm{d}f \, \exp\left[-\frac{Nf}{T} + N s_c(f, T)\right]. \tag{13}$$

For large N, one can as usual perform a saddle-point estimate of the integral, which fixes the dominant value of f, $f^*(T)$ that obeys:

$$T \left.\frac{\partial s_c(f, T)}{\partial f}\right|_{f=f^*(T)} = 1. \tag{14}$$

The temperature dependent complexity is in fact defined by: $s_c(T) \equiv s_c(f^*(T), T)$. The free energy of the system is $f_p(T) = f^* - T s_c(T)$. The typical shape of the configurational entropy as a function of f and a graphic solution of Eq. (13) are plotted in Figure 8. The analysis of the configurational entropy, or complexity $s_c(T)$, reveals that there is a temperature T_K below which $s_c(T)$ vanishes and that $s_c(T)$ increases by increasing the temperature above T_K. There exists a second, higher temperature that we call $T_d > T_K$ (for reasons that will become clear below) above which $s_c(T)$ drops discontinuously to zero again. There is just one minimum[13] above T_d and it corresponds to the homogeneous density profile of the high temperature liquid. The situation below T_K and above T_d is very different. At these two temperatures the part of the free-energy landscape relevant for the thermodynamics change drastically in two very different ways. At T_d the homogeneous liquid state fragments in an exponential number of states, or minima. At T_K the number of minima is no more exponential in the system size, $s_c(T < T_K) = 0$.

---

[13] Actually things are slightly more complicated than this but this is irrelevant for the present discussion.

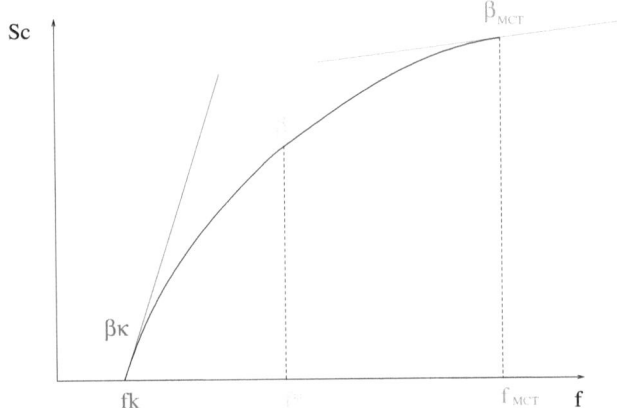

FIGURE 8. Typical shape of the configurational entropy, Σ, as a function of free energy density, f in the range $T_k < T < T_d$ for random first-order landscapes. A graphic solution of Eq. (13) is obtained by finding the value of f at which the slope of the curve is $\beta = 1/T$. Note that s_c is also a function of temperature, so this curve in fact changes with T. T_d is indicated as T_{MCT} for the reasons explained in the text.

Surprisingly $f_p(T)$ is not singular at T_d. This is one of the most unexpected results emerging from the analytical solution. It suggests that at T_d the liquid state fractures in an exponential number of amorphous states and that this transition is only a dynamical phenomenon with no consequences on the thermodynamics. At T_K instead a thermodynamic phase transition takes place since the contribution to the entropy coming from the configurational entropy disappears[14] and, hence, the specific heat makes a jump downward.

In order to understand what's going on at T_d one has to study the dynamics. Such an analysis shows that a typical time dependent correlation function, e.g., $\sum_i \langle n_i(t)n_i(0) \rangle$, develops a plateau as a function of time, t, and has a shape which is very much reminiscent of experimental curves, like Figure 2. Within mean-field theory the barriers between states diverge with N and hence at T_d there is a purely dynamical ergodicity breaking[15].

In summary, the mean-field analysis of the free energy landscape leads to many interesting results. There is a first temperature T_d at which many different amorphous states appear and the "liquid state" fragments into all of them below T_d. A detailed analysis close to T_d shows that they disappear discontinuously: they melt because of thermal fluctuations in a spinodal-like way, which means that they

---

[14]The configurational entropy vanishes linearly at T_K.

[15]This result can be obtained in full detail for mean-field disordered systems only. For more realistic but still mean-field systems, one can either perform a Landau like expansion or numerical simulations. Both strongly support the existence of the dynamical transition discussed in the main text.

lose their stability. Several arguments suggest, and we shall show it directly later on, that all the states we found between T_d and T_K become metastable in finite dimension. As a consequence, the temperature T_d is expected to become a crossover in finite dimension. Instead, at T_K, a thermodynamic transition at which the configurational entropy vanishes takes place. This would lead to a downward jump of the specific heat, to a thermodynamic phase transition and hence, generically, to the divergence of the relaxation time. Clearly, we are finding several features displayed by experiments:

(1) the existence of metastable states that make the dynamics slow down below some temperature,

(2) a possible ideal glass transition where the configurational entropy and the relaxation time diverge.

However, a complete explanation is still lacking. At this point it is not clear yet how the relaxation time diverges and why this is related to the vanishing of the configurational entropy.

It is important to stress that these results are general in the sense that they emerge generically in mean-field treatments of models of glasses, both in the case of liquids approaching the glass transition and hard spheres approaching the glass-jamming transition. However, in practice, performing the analysis of the free energy landscape is already technically quite difficult for lattice glass models on a Bethe lattice. As a consequence, more realistic continuum models of interacting particles cannot be analyzed in full detail. Some numerical and analytical studies of the density functional of more realistic models, such hard spheres, have been performed though. They have indeed revealed that amorphous minima emerge at low energy and high density. In order to compute approximatively the properties of the metastable states and the configurational entropy one has to use different tools. These are explained below and consist in a replica theory for models without quenched disorder.

7.2. Boundary pinning field and replicas

In the following we shall describe an approach based on the replica method which, although more abstract than the one described above, has the clear advantage that analytical mean-field computations for finite-dimensional systems, even realistic ones such as hard spheres or binary Lennard-Jones mixtures, become feasible, although quite involved. Readers allergic to replica or not interested in the formalism can skip this section without any harm: it is not necessary to understand the rest of the paper.

To understand how replicas come about in a model without disorder, let us assume that the system is in a regime of temperatures where there are indeed many very long-lived metastable states and that the Gibbs-Boltzmann measure is distributed over all of them, as found in the previous section. In order to study the statistical property of a typical metastable state and the number of such states, we focus on a very large cavity of radius R, carved in an otherwise infinite (or much larger) system. The basic idea is to apply a suitable boundary external field in an attempt to pin the system in one of the possible metastable states[16]. Contrary to simple cases, e.g., the ferromagnetic transition for which a positive or negative magnetic field selects states, the external "field", or analogously the boundary condition one has to impose to select a given amorphous state, is as unpredictable as the state one wants to select. To overcome this difficulty one can take an equilibrium configuration α, freeze the position of all particles outside the cavity, and use this as a boundary condition. If the system is a thermodynamic glass characterized by many metastable states, then this boundary condition should force the system inside the cavity to be in the same metastable state as the equilibrium configuration α.

Concretely, the procedure consists in computing the cavity partition function, for a fixed α:

$$Z_\alpha(R) = \sum_{\mathcal{C}} \exp(-\beta H(\mathcal{C})) \delta(q^{\text{out}}(\mathcal{C}, \mathcal{C}_\alpha) - 1) \quad, \tag{15}$$

where $q^{\text{out}}(\mathcal{C}, \mathcal{C}_\alpha)$ is a suitably defined overlap that measures the similarity between density configuration $\mathcal{C}$ and that of the α state in the space outside the cavity. When the overlap equals one, the two configurations are the same outside the cavity. In the large R limit, the intensive free energy of the metastable state, obtained by taking the logarithm of the partition function, is expected to be self-averaging and independent of $\mathcal{C}_\alpha$. Physically, this means that the overwhelming majority of the metastable states sampled by the equilibrium Boltzmann measure are characterized by the same intensive free energy.

Although, we started from a problem without quenched disorder, we find that the analysis of the metastable states leads us to a problem where the configuration $\mathcal{C}_\alpha$ plays the role of a (self-induced) quenched disorder. In order to proceed further and compute the intensive free energy of a typical metastable state we have therefore to average over $\mathcal{C}_\alpha$ with the Boltzmann weight $\exp(-\beta H(\mathcal{C}_\alpha))$. As usual for a quenched disorder problem, one can make use of the replica trick:

$$\langle \ln Z_\alpha \rangle_\alpha = \lim_{m \to 1} \frac{\ln \langle Z_\alpha^{m-1} \rangle_\alpha}{m-1}. \tag{16}$$

In order to compute the average $\langle Z_\alpha^{m-1} \rangle_\alpha$ one can introduce replicated configurations and write:

$$\langle Z_\alpha^{m-1} \rangle_\alpha = \frac{\sum_{\mathcal{C}_\alpha; \mathcal{C}_1 \cdots \mathcal{C}_{m-1}} \exp(-\beta H(\mathcal{C}_\alpha)) \prod_{a=1}^{m-1} [\exp(-\beta H(\mathcal{C}_a)) \delta(q^{\text{out}}(\mathcal{C}_\alpha, \mathcal{C}_a) - 1)]}{\sum_{\mathcal{C}_\alpha} \exp(-\beta H(\mathcal{C}_\alpha))}.$$
$$\tag{17}$$

As usual with replicas one computes the above sum for integer and positive values of $m - 1$ and then makes an analytical continuation to make the $m \to 1$ limit. It is important to notice that the average in the numerator of the previous expression can be rewritten as the partition function of m replicas constrained to be identical

---

[16] Our presentation is different from that of the original papers but based on the same ideas.

outside the cavity but free to fluctuate inside, since in the above expression $\mathcal{C}_\alpha$ is no longer different from the other replicas.

Let us denote the logarithm of the partition function of the m constrained replicas as $-\beta F_m$. Once this quantity is known, one can compute the partition function of the large cavity as:

$$\mathcal{F} \equiv -T\langle \ln Z_\alpha \rangle_\alpha = \lim_{m \to 1} \frac{(F_m - F_1)}{m-1} = \left.\frac{\partial F_m}{\partial m}\right|_{m=1}. \tag{18}$$

This gives the free energy of one typical state inside the cavity. Since we are interested in thermodynamic quantities, henceforth we will consider R to be very large. If there are many states, i.e., an exponential number in the size of the system, then the free energy of the cavity $\mathcal{F}$ may be different from the unconstrained one F_1. This can be seen by rewriting the replicated partition function as a sum over all states with their Boltzmann weight. If the constraint is strong enough to force the replicas $a = 1, \ldots, m-1$ to fall into the same state as α itself, then:

$$\begin{aligned}
\frac{F_m}{N} &= -T \frac{\ln \sum_\alpha e^{-\beta f_\alpha mN}}{N} \\
&= -T \frac{\ln \int df \exp(N[-\beta f m + s_c(f,T)])}{N} = f^* m - T s_c(T), \tag{19}
\end{aligned}$$

where f^* is the free energy density that maximizes the argument of the exponential, $s_c(T) = s_c(f^*, T)$ and N is the number of particles inside the cavity. Using Eq. (18), one immediately finds that $f^* = \mathcal{F}/N$ is the intensive free energy of a typical metastable state. But the free energy of the system without constraint is $F_1 = f^* - T s_c(T)$ that contains the configurational entropy contribution. The replica method allows one to obtain both quantities, which yields the configurational entropy:

$$s_c(T) = -\beta F_1 + \beta \left.\frac{\partial F_m}{\partial m}\right|_{m=1} = \beta \left.\frac{\partial}{\partial m}\left[\frac{F_m}{m}\right]\right|_{m=1}. \tag{20}$$

Hence, we have found that computing the statistical properties of metastable states reduces to the computation of the thermodynamics of $m \to 1$ replicas with the constraint that the overlap outside a spherical cavity of radius R is equal to 1. In practice one has to do a computation for m replicas and take the space dependent overlap $q_{a,b}(r)$ between them in the bulk of the cavity as an order parameter. Hence, one has to compute as accurately as possible the free energy as a function of the overlap $q_{a,b}(r)$, and then find the stationary points.

One always finds a trivial solution with uncoupled replicas, i.e., $q_{a,b}(r) = 0$ for $a \neq b$. This is expected since if it was not for the boundary condition the replicas would be indeed completely uncoupled. If this is the only solution then $F_m = mF_1$, and $s_c = 0$ as it should be in a case without metastable states. One has therefore to inspect whether another solution exists. If it is the case then the constraint outside the cavity plays the role of a boundary condition for $q_{a,b}(r)$ and selects this coupled replica solution.

The approach outlined above is the starting point for several quantitative mean-field investigations of realistic glass-formers [23, 24, 20]. It allows one to compute the property of typical metastable states, the configurational entropy and the characteristic temperatures T_K and T_d for realistic systems. The strategy consists in obtaining an expression, as accurate as possible, of the m-replica free energy as a function of the overlap $q_{a,b}$ and then in studying its stationary points and in computing all the physically interesting quantities. Actually, some successful approaches are close in spirit to what I described above but quite different in the details [23].

8. Third dialogue

Cleverus: There are several things I find strange in the mean-field theory you explained before. Why do you sum over all states in Eq. (13) if there are infinite barriers between them, as you have stated?

Salviati: The reason lies in what we think is the generalization of RFOT in finite dimension beyond the mean-field approximation. It is clear that finding so many states with infinite lifetime is an artifact of mean-field theory. For instance, it is impossible for a system to display two thermodynamic states with different *intensive* free energies, because if the system sits in the highest free-energy one, it will eventually nucleate the other, thus showing that the initial one didn't have an infinite life-time. We expect that all the states found in mean-field theory between T_K and T_d will acquire a finite life time in a correct treatment. That is the reason why we sum over all of them in Eq. (13).

Cleverus: OK, well, you will have then to explain me what the dynamics between metastable states is, because there lies the final explanation of the glass transition if RFOT is right.

Salviati: Yes, sure.

Cleverus: But, before, I have still some questions on RFOT: part I. You seemed to be happy of finding T_K and T_d. T_K, I understand, based on what you explained before, but what about T_d?

Salviati: Yes, I didn't have time to discuss this point, which is interesting. The physical properties of the dynamical transition taking place at T_d coincide with the ones found for the so-called Mode Coupling Transition Theory (MCT) of glasses. This is a self-consistent approach for the dynamics of dense and low temperature liquids that was developed by Gotze and collaborators. It has been already tested a lot in simulations and experiments and, indeed, it seems that there is a characteristic temperature where a dynamical cross-over takes place and that this is rather well described by MCT.

Cleverus: Really? I didn't have this impression reading some papers about MCT.

Salviati: You are right, maybe I'm a bit biased. The problem is that the dynamical transition becomes just a cross-over. And a cross-over is difficult to test in a clear cut way. Furthermore, even theoretically, we don't understand well the properties of this cross-over yet.

Cleverus: I have a final question about grand-mothers.

Salviati: Uh?

Cleverys: Yes! In the papers I brought with me, I often read strange sentences stating that even a grand-mother would have a glass transition with your approximations. This apparently shows two things: first, you guys don't respect much grand-mothers, which is bad, and second, that what you find is just due to the approximations you have done.

Salviati: Well, yes and no. No, we respect grand-mothers, this was just a joke. Yes, what we find is due to the approximations but we think that a correct treatment would change somehow the results without jeopardizing the whole scenario.

Cleverus: Is this wishful thinking or is it based on some concrete analysis?

Salviati: In the last years indeed we have found ways to perform concrete analysis and even test our ideas in numerical simulations. I can explain if you want.

Cleverus: Sure! Keep in mind that the things I would like to know are: (1) what happens in real space within the RFOT scenario. You discussed a lot the importance of understanding the glass transition in real space in Section 5, but in your previous sections about RFOT real space was completely out of the game. (2) Why and how the relaxation time diverges at T_K within RFOT?

9. The Random First-Order Transition Theory Part II: Beyond mean-field theory and the real space description

9.1. The Mosaic State

We have discussed already that the multiple states found in mean-field theory must become metastable, i.e., with a finite lifetime, in a correct treatment. As Cleverus correctly pointed out, understanding how this comes about and how the dynamics between metastable states takes place is central for the application of RFOT to real glass-formers and hard spheres systems.

In a pioneering work, Kirkpatrick, Thirumalai and Wolynes [25] proposed that liquids must become a mosaic of mean-field states with a typical "tile" size of the order of $l^* \propto 1/(T - T_K)$. As before, we shall not reproduce the original arguments but present more recent (and at least to our eyes clearer) ones [26].

Consider the following *Gedanken* experiment. Suppose we could identify one of the exponentially numerous TAP states relevant at a given temperature T, which we call α, and characterize the average position of all the particles in that state. We shall first establish that there exists a length scale above which the assumption that this TAP state has an infinite lifetime is inconsistent. In order to do this, we freeze the motion of all particles outside a spherical cavity of radius R and focus on the *thermodynamics* of the particles inside the sphere, $\mathcal{S}(R)$, that are free to move but are subject to the boundary conditions imposed by the frozen particles outside the sphere. Because of the 'pinning' field imposed by these frozen particles, some configurations inside $\mathcal{S}(R)$ are particularly favored energetically. When $s_c(T)R^d$ is much larger than unity there are many metastable states accessible to the particles in the cavity. The boundary condition imposed by the external particles, frozen in state α, act as a random boundary field for all other metastable states except α itself, for which these boundary conditions perfectly match. Any other metastable state γ has a positive mismatch energy. We assume this interface energy can be written as $\Upsilon_0 R^\theta$. We first imagine that we wait long enough so that the cavity embedded in state α is fully equilibrated. The partition function Z_α can then be decomposed into two contributions:

$$Z_\alpha(R,T) = \exp[-\Omega_d R^d \frac{f_\alpha}{T}] + \sum_{\gamma \neq \alpha} \exp\left[-\Omega_d R^d \frac{f_\gamma}{T} - \frac{\Upsilon_0 R^\theta}{T}\right] \qquad (21)$$

$$\approx \exp[-\Omega_d R^d \frac{f_\alpha}{T}] + \int_{f_{\min}}^{f_{\max}} df \exp\left[\frac{(Ts_c(f,T)-f)\Omega_d R^d - \Upsilon_0 R^\theta}{T}\right],$$

where f_γ is the excess free energy per unit volume of state γ, Ω_d the volume of the sphere with unit radius in dimension d[17]. We focus on the case T close to T_K where R is large, allowing one to invoke saddle point arguments. We focus on a typical state α at that temperature, i.e., a state with the free energy f^* that dominates the integral over f above, such that $T \partial s_c/\partial f = 1$. The partition function of the cavity embedded in the α state becomes independent of α and reads:

$$Z(R,T) \approx \exp\left[-\Omega_d R^d \frac{f^*}{T}\right]\left(1 + \exp\left[\Omega_d R^d s_c(T) - \frac{\Upsilon_0 R^\theta}{T}\right]\right). \qquad (22)$$

The above expression is central to our argument. It points out the existence of a cross-over length ℓ^*:

$$\ell^* = \left(\frac{\theta \Upsilon_0}{d \Omega_d T s_c(T)}\right)^{\frac{1}{d-\theta}}. \qquad (23)$$

When R is smaller than ℓ^* but still large, the second term is exponentially small even if an exponentially large number of terms contribute. The mismatch energy dominates and the state α favored by the boundary conditions is the most probable state, even if the particles inside the cavity are free to move. In this sense, the cavity is in a glass phase, where only one (or a few) amorphous configurations, selected by the boundary conditions, are relevant. When $R > \ell^*$, on the other hand, the second term becomes overwhelming. There are *so many* other states to explore that it becomes very improbable to observe the α state. There is "entropic melting" of the cavity. Note that approaching T_K, $s_c(T) \to 0$ and the length ℓ^* diverges.

The main conclusion is that a TAP state does not make sense on scales larger than l^*. The arguments above strongly suggest that if one starts from the

---

[17]Here and below, all lengths are in units of the interparticle distance a.

mean-field free energy and adds all fluctuations needed to recover the correct result, then one would obtain the following: For lengths $R < \ell^*$ there is a set of boundary conditions indexed by α such that the true free energy – not the one computed within the mean-field approximation – has one different minimum for each boundary condition. Each minimum corresponds to a density profile which is very close to the one, called α, obtained by analyzing the mean-field free energy. Instead, for $R \gg \ell^*$, there is just one minimum, or analogously, the density profile at the minimum is just the flat liquid one and is insensitive to the boundary conditions, whatever they are.

One therefore identifies the scale ℓ^* as the one over which the liquid can be considered as formed by independent pieces or sub-systems[18]. A sketchy representation of this state is a *mosaic* composed by tiles which are the different TAP states that exist on the scale ℓ^*. The free-energy of the liquid is therefore, up to a subleading contribution:

$$f_{\text{liq}} \approx f^*(T) - T s_c(T), \tag{24}$$

which is formally identical to the mean-field result but very different in terms of the real space interpretation. Approaching T_K the mosaic tiles increase in size and at T_K there is a spontaneous symmetry breaking of translation invariance toward states with amorphous long range order.

Let us now discuss the dynamics. Since the picture we ended up with is one where the liquid is composed by a mosaic of different states on the scale ℓ^*, it is natural to expect that the relaxation time of the system is the relaxation time, $\tau(\ell^*)$, of a finite size region of the system of size ℓ^*. Contrary to mean-field theory, barriers are now finite, hence T_d is no longer a true transition. Focusing for the time being on the dynamical evolution close to T_K, it is natural to assume that the dynamical process leading to relaxation of the cavity is thermal activation over energy barriers which are supposed to grow with size as ξ^ψ, as in disordered systems. One predicts finally that:

$$\log\left(\frac{\tau_\alpha}{\tau_0}\right) = c\frac{\Upsilon_0}{T}(\ell^*)^\psi = c\frac{\Upsilon_0}{T}\left(\frac{\Upsilon_0}{T s_c(T)}\right)^{\psi/(3-\theta)}, \tag{25}$$

where c is a constant. Kirkpatrick, Thirumalai and Wolynes argued that $\psi = \theta = 3/2$, and hence $\psi/(3-\theta) = 1$, thus recovering the Adam-Gibbs law, at least close enough to T_K. This is one of the main results of RFOT: the decrease of the configurational entropy and the increase of the effective barrier are linked together: the latter has to diverge when the former vanishes. Furthermore, with $\psi/(3-\theta) = 1$ one indeed recovers the Vogel-Fulcher law and obtains the identity $T_0 = T_K$.

This is certainly very impressive. However, we are still far from having this theory and, hence, these results do not lie on a firm ground. In fact, the previous argument is a phenomenological one. However, it has been backed up by microscopic computations based on the replica formalism discussed in a previous

---

[18]Using the concept of "generalized rigidity" of Anderson, one would say that above ℓ^* the system is no longer rigid.

section [27, 28]. These analyses have indeed confirmed the existence of the length ℓ^*. Technically, they found an instability length of the homogenous strongly coupled replica solution using instanton techniques. However, the exponent θ that has been found is equal to 2 and not $3/2$. As for the exponent ψ, there is no reliable computation yet. We are at a stage where we don't know how to compute reliably the exponents and, it is not clear yet that the victory claimed by Wolynes and collaborators is the final word. More work is ahead.

9.2. Dynamics close to T_d

We have found that in mean-field theory, amorphous states lose their stability in a spinodal-like way close to T_d. Approaching T_d from below the frozen part of density fluctuations, called Edwards-Anderson parameters in the spin-glass jargon, does not go to zero although the states become unstable just above T_d. This means that, approaching T_d from above, dynamical two-point correlation functions must show a plateau whose value converges at T_d toward the Edwards-Anderson parameter of the TAP states. Above T_d, one expects a behavior of the type found in experiments, see Figure 1: first a relaxation toward a plateau and then a relaxation away from it.

As we have shown, TAP states are unstable in a correct – beyond mean-field- treatment. As a consequence, at T_d there is no dynamical transition, but instead possibly a cross-over. Understanding this cross-over is still an open problem, but one can already obtain many interesting results on the real space dynamics pretending that the transition is real. Of course, it is a crucial issue understanding how much these results persist in a correct treatment where the transition is replaced by a cross-over. In the following we shall list some of the main properties of the dynamics close to T_d that one can obtain within the mean-field approximation. Note that, contrary, to what happens below T_d, where mean-field results on the dynamics are completely changed in a correct treatment, we expect that this shouldn't be the case above T_d at least not too close to T_d.

First, mean-field theory leads to specific predictions on the form of the dynamical correlation function, which has been tested during the last decades, because they coincide with the ones obtained from MCT. They are qualitatively correct and quantitatively in rather good agreement both for models of glass-formers and for hard sphere systems. Unfortunately, the existence of the cross-over makes any precise test difficult if not impossible.

Furthermore mean-field theory correctly reproduces qualitatively the vast majority of the phenomena discussed in Section 5 under the name of dynamical heterogeneity and dynamical correlations. One obtains from first principle computations a diverging χ_4 and a diverging dynamical correlation length at T_d. Physically, this is due to the fact that TAP states become unstable at T_d because of the emergence of amorphous soft modes that make the system yield above T_d. The dynamics above T_d takes place mainly along these soft modes and is therefore correlated on a large scale. The quantitative predictions for dynamical correlations and dynamical heterogeneity are under current investigation. The evolution of χ_4 is quite in agreement with simulation results, although of course it does not diverge at T_d since the transition is avoided. Actually, one finds results whose deviations remain very close to T_d. Instead the agreement for the growing correlation length is under debate because it appears very difficult to extract reliably the precise values of ξ_d in simulations.

10. Fourth and final dialogue

Cleverus: I see that you still have to wave many hands, but I understand your theory – more or less. There are many loose ends though. You have work to do!

Salviati: Yes, right.

Cleverus: Your Gedanken experiment is nice but couldn't you test it at least in a simulation?

Salviati: This has been done actually in [29] for a model of glass-forming liquids: the authors froze a liquid equilibrium configuration and re-equilibrated the cavity. They found that after re-equilibration the configuration at the center of the cavity becomes independent of boundary conditions only for radii larger than a typical length ℓ^*, which increases approaching the glass transition. They also found that the relaxation time changes from the bulk behavior for radii smaller than ℓ^*.

Cleverus: Interesting. This indicates that maybe you are on the right track. It also shows that, again, a complete local explanation of the dynamical slowing down is completely hopeless. But what about other theories? I guess that their advocates wouldn't be so satisfied with RFOT as you are.

Salviati: You bet! They wouldn't certainly be. There are indeed other theories whose advocates also claim at least partial victory. I think we are still a bit far from the final solution. Our current most important challenge is to nail down what is the closest starting point to the Theory of the glass transition. I believe this is RFOT for several reasons that I partially try to explain before: In particular it's what it comes out naturally from mean-field theory and when one tries to correct mean-field theory one ends up with results that are physically sound.

Cleverus: Yes, but for example you don't know how to compute the exponents θ and ψ. Maybe a correct computation would show that $\theta = 0$, so states are unstable, or that $\psi/(3 - \theta)$ is very far from 1, so bye bye RFOT!

Salviati: Yes, dear Cleverus, it's indeed a possibility. However, look at the good side of it: there is a clear route to follow, ahead of us, in order to prove whether the theory is right or wrong and many highly non-trivial tests to do to reinforce and eventually validate or disprove RFOT. Note, for example, that the simulation of [29] was indeed already a very important test to pass.

Cleverus: I see ... you are viciously implying that this is not the case for the other theories.

Salviati: Nah, not all theories at least. Anyway, I am confident that in a few years from now we will be at a stage where it will be clear beyond any reasonable doubt which theory is the good starting point.

Cleverus: All the discussion about the dynamical transition looked fishy: you seems to claim victory saying that it works, and when it does not, you say it is because of the cross-over. Isn't it a bit too easy?

Salviati: Well, yes and no. No, there are some qualitative results of the mean-field theory that are really non-trivial and that one finds in simulations of glass-formers. These results just come straight from first principle computations, they could have been very different, you know. For example, no growing dynamic correlations or growing in a weird way and then, bye bye, mean-field theory. However, yes, you are right, the lack of a good description of the cross-over close to T_d is a weak point.

Cleverus: But, physically, you said that it should be related to the emergence of soft modes in the states and that the dynamical relaxation should mainly take place along them. This is something it should be possible to test in simulations.

Salviati: Actually, yes! This has been done for example in [30] for glass-formers and also in [31] for hard spheres.

Cleverus: OK. So, finally, tell me your view on glassy relax – not the balanced one you write in papers but the wild guess.

Salviati: Mhmm, ok. I think that close to T_d the dynamics is a mix of activation combined with relaxation along the soft modes. There, the dynamics is characterized by small clusters of particles, which relax by activation and cooperatively. And this triggers an avalanche process along soft modes or regions that are soft. Actually, some of what I am saying has been found in simulations [10] of glass-formers and in experiments of granular systems [12]. Decreasing the temperature, the importance of the soft modes decreases, cooperative clusters increase in size and the relaxation time is determined by some variants of the RFOT argument described before.

Cleverus: Two final questions. From the point of view of RFOT, hard sphere colloids and molecular liquids display the very same transition. But what about granular systems, they are phenomenologically very similar but you cannot describe them in terms of Boltzmann equilibrium, they are in an out of equilibrium steady state.

Salviati: You are right. This is puzzling and very fascinating. Could it be that despite the important physical differences, the glass transitions of liquids and grains are driven by the same underlying critical phenomenon? One experiment I would love to see is the Gendanken one for a granular system. Maybe, it would show that the very same kind of subtle amorphous *static* correlations develops also in a granular system despite the stationary distribution is not Boltzmann.

Cleverus: Final question or comment: you swept the crystal problem under the rug. But in general there is an underlying crystal, so the super-cooled liquid state is not a true equilibrium state and the glass transition cannot be a true phase transition.

Salviati: Com'on Cleverus ... I got the same comment from Simplicius.

Cleverus: OK, right, it was just to tease you. Thank you very much for all the explanations. No way, I could have understood all this alone during my sailing trip. It was good to see you. By the way, you cited a lot Anderson, but you never quoted the last sentence of his famous article on glasses: "The solution of the more important and puzzling glass problem may also have a substantial intellectual spin-off. Whether it will help make better glass is questionable." Bye Bye.

Acknowledgment

I thank all my glassy collaborators, in particular Jean-Philippe Bouchaud, my long-standing collaborator on glasses. Our last five years of research together were a great adventure. I also would like to acknowledge that some parts of this paper have been largely "inspired" by the reviews I have written with L. Berthier and J.-P. Bouchaud. I thank both of them for the patience they had in writing those reviews with me.

Some References

Since this is meant to be an informal introduction to the glass and jamming transitions, in the main text I almost didn't cite any paper besides very few random exceptions. I owe an apology to all my colleagues that are not finding their works cited. I suggest below few formal reviews for readers which want to know more about the topics treated in this paper.

Sections 1–5

L. Berthier and G. Biroli, *A theoretical perspective on the glass transition and nonequilibrium phenomena in disordered materials*, arxiv.org 1011.2578, submitted to Review of Modern Physics.
P.G. Debenedetti and F.H. Stillinger, Nature **410**, 259 (2001).
The book "Dynamical heterogeneities in glasses, colloids and granular materials" to be published in 2011 by Oxford University Press, see http://w3.lcvn.univ-montp2.fr/~lucacip/DH_book.htm.

Section 7–9

T.R. Kirkpatrick, D. Thirumalai, and P.G. Wolynes, Phys. Rev. A **40**, 1045 (1989).
J.-P. Bouchaud and G. Biroli, J. Chem. Phys. **121**, 7347 (2004).
A. Cavagna, *Supercooled liquids for pedestrians*, Physics Reports 476, 51 (2009).
G. Biroli and J.-P. Bouchaud, arXiv:0912.2542.

References

[1] G. Galilei, *Dialogue Concerning the Two Chief World Systems*.
[2] S. Coleman, *Aspects of Symmetry*, Cambridge University Press, 1985.
[3] P.G. Debenedetti and F.H. Stillinger, Nature **410** (2001), 259.
[4] R. Richert and C.A. Angell, J. Chem. Phys. **108** (1998), 9016.
[5] H. Tanaka, Phys. Rev. Lett. **90**, 055701 (2003).
[6] C.A. Angell and S. Borick, J. Non-Cryst. Solids **307**, 393 (2002).
[7] J. Dyre, Rev. Mod. Phys. **78**, 953 (2006).
[8] M. Wyart, Phys. Rev. Lett. **104**, 095901 (2010).

[9] J. Wuttke, W. Petry, and S. Pouget, J. Chem. Phys. **105** (1996), 5177.

[10] R. Candelier, PhD Thesis, Univ. Paris VI (2009).

[11] E.R. Weeks, J.C. Crocker, A.C. Levitt, A. Schofield, and D.A. Weitz, Science **287** (2000), 627.

[12] R. Candelier, O. Dauchot, and G. Biroli, Phys. Rev. Lett, **102** (2009), 088001.

[13] L. Berthier, G. Biroli, J.-P. Bouchaud, R.L. Jack, arXiv:1009.4765

[14] L. Berthier and G. Biroli, *A Statistical Mechanics Perspective on Glasses and Aging* Encyclopedia of Complexity and Systems Science, Springer (2009).

[15] A.S. Keys, A.R. Abate, S.C. Glotzer, and D.J. Durian, Nature Physics **3** (2007), 260.

[16] L. Berthier, G. Biroli, J.-P. Bouchaud, L. Cipelletti, D. El Masri, D. L'Hôte, F. Ladieu, and M. Pierno, Science **310** (2005), 1797.

[17] C. Dalle-Ferrier, C. Thibierge, C. Alba-Simionesco, L. Berthier, G. Biroli, J.-P. Bouchaud, F. Ladieu, D. L'Hôte, and G. Tarjus, Phys. Rev. E **76** (2007), 041510.

[18] A. Montanari, G. Semerjian, J. Stat. Phys. **125** (2006), 23.

[19] F. Krzakala, A. Montanari, F. Ricci-Tersenghi, G. Semerjian, L. Zdeborova, Proc. Natl. Acad. Sci. **104** (2007), 10318.

[20] G. Parisi and F. Zamponi, arXiv:0802.2180.

[21] L. Santen and W. Krauth, Nature **405** (2000), 550.

[22] L. Berthier and T.A. Witten, Phys. Rev. E **80** (2009), 021502.

[23] M. Mézard and G. Parisi, Phys. Rev. Lett. **82** (1999), 747.

[24] S. Franz, M. Cardenas, G. Parisi J.Phys. A: Math. Gen. **31** (1998), L163.

[25] T.R. Kirkpatrick, D. Thirumalai, and P.G. Wolynes, Phys. Rev. A **40** (1989), 1045.

[26] J.-P. Bouchaud and G. Biroli, J. Chem. Phys. **121** (2004), 7347.

[27] S. Franz, A. Montanari J. Phys. A: Math. and Theor. **40** (2007), F251.

[28] M. Dzero, J. Schmalian, and P.G. Wolynes, Phys. Rev. B **72** (2005), 100201.

[29] G. Biroli, J.-P. Bouchaud, A. Cavagna, T.S. Grigera, P. Verrocchio, Nature Physics **4** (2008), 771. C. Cammarota, A. Cavagna, G. Gradenigo, T.S. Grigera, P. Verrocchio, arXiv:0904.1522; Chiara Cammarota, Andrea Cavagna, Giacomo Gradenigo, Tomas S. Grigera, Paolo Verrocchio, arXiv:0906.3868.

[30] A. Widmer-Cooper, H. Perry, P. Harrowell and D.R. Reichman, Nature Physics **4** (2008), 711.

[31] C. Brito, M. Wyart, J. Chem. Phys. **131** (2009), 024504.

Giulio Biroli
Institut de Physique Théorique
Orme des Merisiers
CEA Saclay
F-91191 Gif-sur-Yvette Cedex, France
e-mail: `giulio.biroli@cea.fr`

Glasses and Grains, 77–109
© 2011 Springer Basel AG

Granular Flows

Yoël Forterre and Olivier Pouliquen

Abstract. Those who have played with sand on the beach or with sugar in their kitchen are aware that a collection of solid grains can behave macroscopically like a liquid flow. However, the description of this peculiar fluid still represents a challenge due to the lack of constitutive laws able to describe the rich phenomenology observed. In this paper we review the properties of dry granular flows and we present recent advances in our understanding of their rheological behavior. The success and limits of a simple visco-plastic model recently developed is presented. In a second part of the paper, we go beyond the simple and ideal situation where the material is made of grains having the same size and interacting by contact interactions only. We present studies on more complex and realistic granular media such as polydispersed media, cohesive granular media or granular pastes made of grains immersed in a liquid. We analyse to what extent the progress made in our description of monodispersed dry granular flows can help us to understand these more complex granular media.

1. Introduction

Sand, gravels, rice, sugar ... Granular matter is everywhere in our everyday life (Figure 1). Strong enough to support buildings, a granular medium can flow like a liquid, for example in an hourglass, or can be transported by the wind to create dunes in the desert. This variety of behaviour represents one of the difficulties of the physics of granular media [45]. Research in this area is motivated by numerous applications encountered in industrial processes and especially in geophysics for the description and prediction of natural hazards like landslides, rock avalanches or pyroclastic flows. However, the recent interest in granular flows is certainly also stimulated by new fundamental questions raised by this peculiar fluid, which shares similarities with other athermal disordered systems such as foam, amorphous solids or emulsions [63, 17] and which exhibits a very rich phenomenology [5].

A granular medium is a collection of macroscopic particles, their size being typically greater than 100 μm. This limitation in size corresponds to a limitation in

FIGURE 1. Examples of granular materials.

the type of interaction between particles. A granular medium means non-brownian particles, which interact solely by friction and collision. For smaller particles, other interactions such as Van der Waals forces or humidity, start to play a role and one enters the world of powder. At even smaller sizes, below 1 μm, thermal agitation is no longer negligible and one enters the world of colloids. Granular media are then a priori simple systems made of solid particles interacting through contact interactions. However, they still resist our understanding and no theoretical framework is available to describe the variety of behaviours observed. One can try to list the difficulties encountered when dealing with granular material.

First, granular media are composed of a large number of particles. A spoon of sugar contains more than one millions of grains, which is larger than what we can compute numerically with ideal spherical particles. There is then a need for a continuous description, trying to define averaged quantities and to model the granular medium as a continuum medium. A large number of particles is not necessarily a serious obstacle, if one consider gases or liquids, for which the number of molecules is much larger than the number of grains in a spoon of sugar. However, in the case of gases or liquids, the presence of thermal agitation allows a proper statistical approach, allowing us to derive macroscopic quantities from microscopic ones. In the case of a granular medium, particles are too large to experience Brownian motion and a statistical average over different configurations is not possible. The system is stuck in metastable states. Granular media are then often qualified as athermal systems [63]. Another difficulty encountered when one tries to apply tools from statistical physics to granular media, is the dissipative nature of the particle interaction. Contact interactions including friction and inelastic collisions are highly nonlinear and dissipative mechanisms. This dissipation at the microscopic level is an important difference with classical systems studied in statistical physics. The continuum description of granular media is also made difficult by the lack of a clear scale separation between the microscopic scale, i.e., the grain size, and the macroscopic scale, i.e., the size of the flow. Typically, when sand flows down on a pile, the flow thickness is about 10 to 20 particle diameters. The physics of granular media shares this difficulty with nanofluidics, when effects of the size of the molecule start to play a role. The last difficulty is the observation that granular media exists under different states [83]. Depending on the way it is handled, a granular material can behave like a solid, a liquid or a gas (Figure 2). Grains can sustain stresses and create a static pile, but can also flow like a liquid in an hourglass, or can create a gas when they are strongly agitated. These different flow regimes can coexist in a single configuration as shown by the flow of beads on a pile shown in Figure 2.

"solid"   "liquid"   "gas"

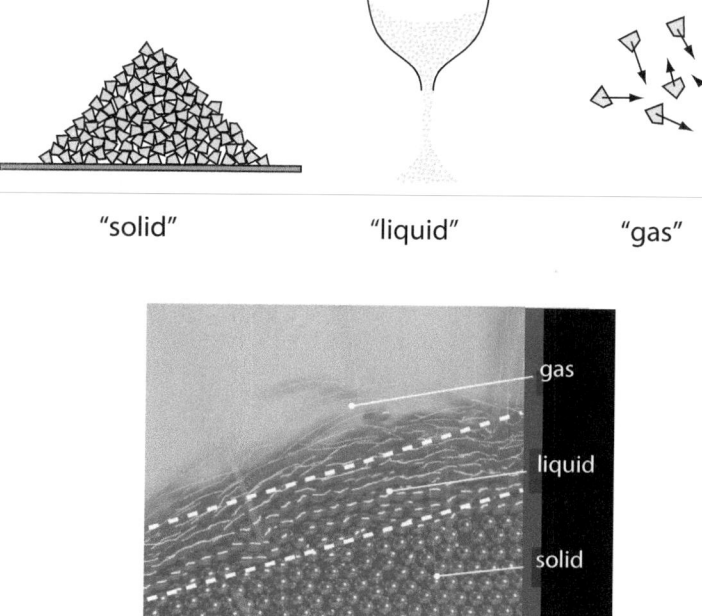

FIGURE 2. Different states of granular media.

In the face of such a complexity, different frameworks have been developed to describe the different flow regimes. The dense quasi-static regime where the deformations are very slow and the particles interact by frictional contacts is usually described using plasticity theories [100, 109, 91]. The gaseous regime where the flow is very rapid and dilute and the particles interact by collision [38] has motivated a lot of work based on the concept of the kinetic theory of gases. The intermediate liquid regime where the material is dense but still flows like a liquid is the less understood regime [34]. In this paper we focus on this last liquid regime, which is most often encountered in applications, and discuss the possibility of a hydrodynamic description of dense granular flows.

2. The granular liquid

Rock avalanches, flows of cereals out of a silo, are typical examples of dense granular flows. In this flow regime, the volume fraction of the material (ratio of the volume occupied by the grains to the total volume) is high and close to the maximum value. The grains interact both by friction and collision through a contact network. From a phenomenological point of view, the material flows like a liquid with peculiar features. To better understand this regime, different flow configurations have been investigated, the most common being sketched in Figure 3. They can be divided in two families: flows confined between walls as in shear cells or silo, and free surface flows like flows down an inclined plane, flows in rotating drums or flows on a pile. Their characteristics in terms of velocity profiles, density profiles and velocity fluctuations are discussed in detail in [36]. Recently, by analogy with classical hydrodynamics problems, more complex flow configurations have been analyzed, such as dam break problems [59, 68], coating-like problems [28, 22], mixing experiments [76], split Couette devices [30], drag problems [55] and instabilities [5].

A recurrent and central question underlying all these studies is the question of the constitutive equations of this peculiar liquid. Dense granular flows belong to the visco-plastic family of materials because of two broad properties. First, a flow threshold exists, although it is expressed in terms of friction instead of a yield stress, as in a classical visco-plastic material. Second, when the material is flowing, shear rate dependence is observed, which gives a viscous-like behavior. In the following section we present recent advances in our understanding of the rheology of dense granular flows. We first discuss the plane shear configuration, which provides the basic ideas allowing us to propose a constitutive law for dense granular flows. The applications to other configurations are discussed and the limits of this simple local rheology is discussed.

2.1. Local rheology

Let us consider a granular material made of particles of diameter d and density ρ_p under a confinement pressure P. The material is confined between two rough

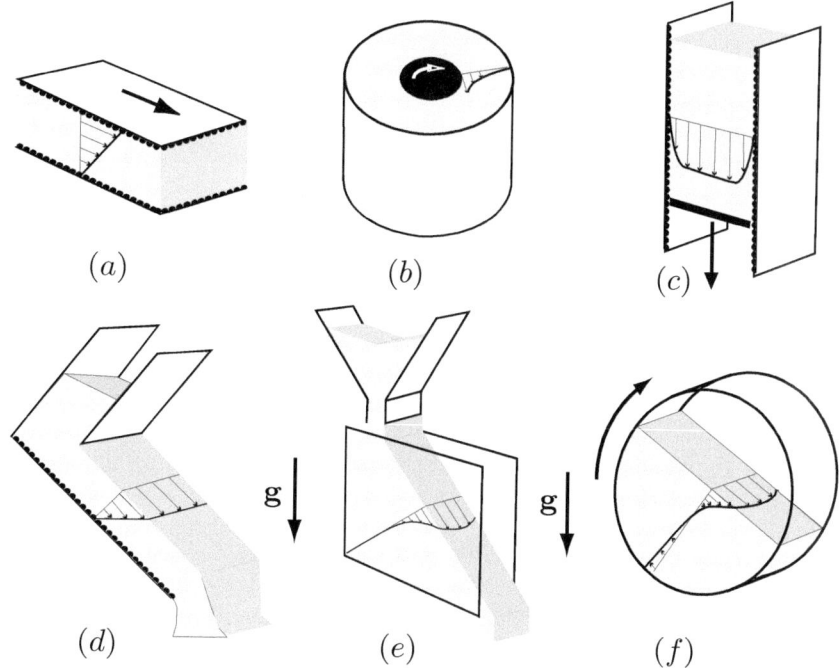

FIGURE 3. Different configurations used to study granular flows. (a)
Plane shear (b) Couette cell (c) vertical silo (d) Inclined plane (e) Heap
flow (f) rotating drum.

plates by a pressure P imposed to the top plate. The material is sheared at a
given shear rate $\dot{\gamma} = V_w/L$ imposed by the relative displacement of the top plate
at a velocity V_w. (Figure 4). In absence of gravity, the force balance implies that
both the shear stress $\tau = \sigma_{xz}$ and the normal stress $P = \sigma_{zz}$ are homogeneous
across the cell. This configuration is then the simplest configuration to study the
rheology of granular flows, namely to study how the shear stress τ and the volume
fraction ϕ vary with the shear rate $\dot{\gamma}$ and the pressure P.

A crucial observation raised by Da cruz et $al.$ [18, 19] and by Lois et $al.$ [64] is
that, in the simple sheared configuration for infinitely rigid particles, dimensional
analysis strongly constrains the stress/shear rate relations [36]. For large systems
($L/d >> 1$), and for rigid particles (the young modulus being much higher than the
confining pressure), the system is controlled by a single dimensionless parameter
called the inertial number:

$$I = \frac{\dot{\gamma}d}{\sqrt{P/\rho_p}}. \tag{1}$$

As a consequence, dimensional analysis imposes that the volume fraction ϕ
is a function of I only, and that the shear stress τ has to be proportional to the

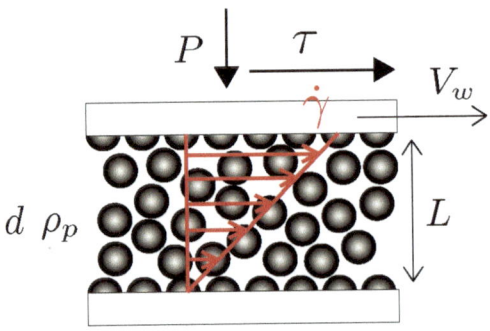

FIGURE 4. Plane shear at constant pressure.

normal stress P, which is the only stress scale of the problem. The constitutive
laws can then be written as follows:

$$\tau = P\mu(I) \quad \text{and} \quad \phi = \phi(I), \tag{2}$$

where $\mu(I)$ is a friction coefficient, which depends on the inertial number. The
shape of the friction coefficient $\mu(I)$ and of the volume fraction $\phi(I)$ are provided
by numerical simulations using discrete element models and by experimental measurements. Figure 5 presents a summary of results coming from different studies
for 2D systems (disks) or 3D (spheres). One observes that the friction coefficient μ
is an increasing function of the inertial number. Friction increases when increasing
the shear rate and/or decreasing the pressure. In the limit of quasi-static flows
$(I- > 0)$ the friction coefficient tends towards a constant. The volume fraction
also varies with I. It starts at a maximum value when $(I- > 0)$ and decreases more
or less linearly with I. It is interesting to note that in the range of inertial number
corresponding to the dense flow regime, the macroscopic friction coefficient $\mu(I)$
and the volume fraction $\phi(I)$ do not depend on microscopic properties of grains.
Changing the coefficient of restitution of the grains, or changing the inter-particle
friction coefficient (as long as it is not zero), does not change the macroscopic
friction [19].

The inertial number appears to be the most important parameter controlling
the rheology of dense granular flows. It can be interpreted in terms of the ratio
between two time scales: a microscopic time scale $d/\sqrt{P/\rho_p}$, which represents the
time it takes for a particle to fall in a hole of size d under the pressure P, and
which gives the typical timescale of rearrangements; and a macroscopic timescale
$1/\dot{\gamma}$ linked to the mean deformation. This interpretation allows to classify more
precisely the different flow regimes. Small I correspond to a quasi-static regime
in the sense that macroscopic deformation is slow compared to microscopic rearrangements, whereas large values of I correspond to rapid flows. The dimensional
analysis tells us that, to switch from quasi-static to inertial regime, one can either

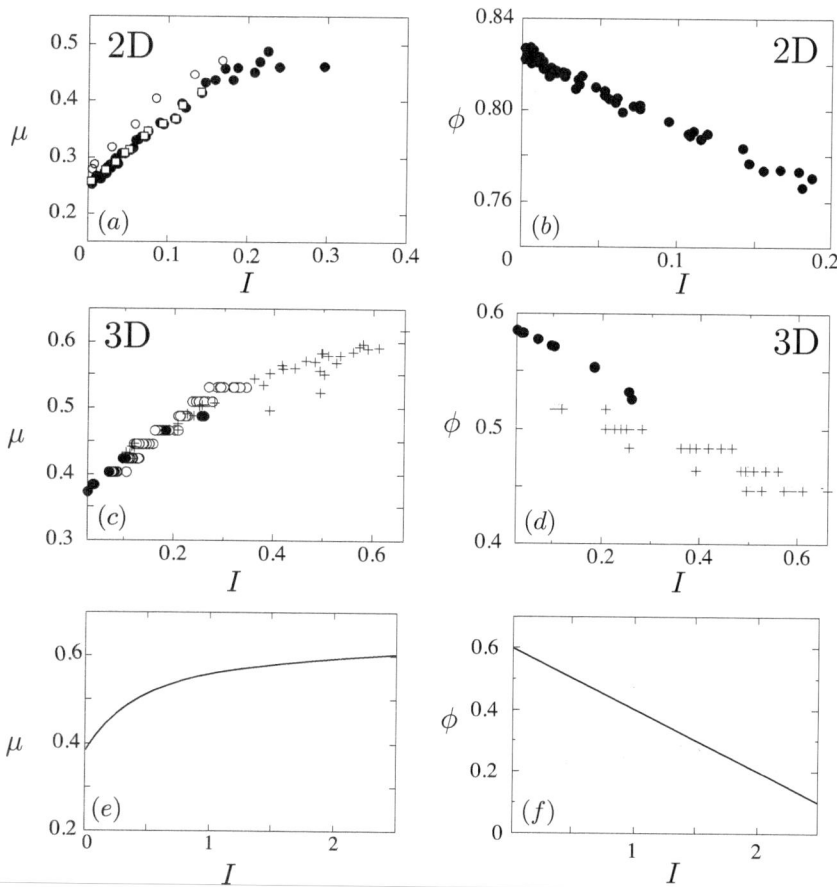

FIGURE 5. Friction law $\mu(I)$ and volume fraction law $(\phi(I))$; (a) (b) for 2D configurations with disks (c) (d) for 3D configurations with spheres; (e) (f) empirical analytical law proposed (Eqs. (3)) data from [80, 96, 64, 102].

increase the shear rate or decrease the pressure. This inertial number is also equivalent to the square root of the Savage number or Coulomb number introduced by some authors as the ratio of collisional stress to total stress [96, 1].

In the plane shear configuration, the velocity profile is linear. It is then tempting to assume that relations (2) obtained in this configuration give the intrinsic rheology of the granular medium. This is true only if the stresses that develop in an inhomogeneous flow are the same as in the plane shear. It is the case if the rheology is local, namely, if the stresses depend only on the local shear rate and on the local pressure. Under the assumption of a local rheology one can then use Eqs.

(2) as constitutive equations. Fitting the experiments and numerical simulations, it is possible to propose analytical expressions for the friction law and the volume fraction law, which can then be used to study other configurations. An example of phenomenological expressions are:

$$\mu(I) = \mu_1 + \frac{\mu_2 - \mu_1}{I_0/I + 1} \qquad \text{and} \qquad \phi = \phi_{\max} + (\phi_{\min} - \phi_{\max})I. \qquad (3)$$

Typical values of the constants obtained for monodispersed glass beads in 3D are: $\mu_1 = \tan 21^o$, $\mu_2 = \tan 33^o$, $I_0 = 0.3$, $\phi_{\max} = 0.6$, $\phi_{\min} = 0.4$. Those functional forms have not been tested for large values of the inertial number I. However, the choice of a friction law that saturates to a finite value μ_2 when I goes to infinity is supported by experiments of steady granular fronts flowing down a slope [81]: at the tip of a front the shear rate goes to infinity, whereas experiments reveal that the slope, and hence the friction coefficient, remains finite. This is consistent with the saturation of $\mu(I)$ to μ_2.

Before discussing applications of this simple phenomenological description of granular flows, it is important to say that Eq. (2) can be generalised to a tensorial form. When written in terms of a scalar like in (2), the rheology can only describe flows sheared in a single direction. In order to describe more complex 3D configurations the friction law has to be written in terms of the shear rate tensor [51]. The simplest way to do so, is to assume that the flow is incompressible, i.e., we neglect variation of volume fraction and that the pressure is isotropic. We also assumed that the shear stress tensor is colinear to the shear rate tensor, as proposed by previous authors [37, 94] and as suggested by numerical simulations [23]. The stress tensor can then be written in terms of an effective viscosity as follows:

$$\sigma_{ij} = -P\delta_{ij} + \tau_{ij} \qquad (4)$$

where P is the isotropic pressure,

$$\tau_{ij} = \eta_{eff}\dot{\gamma}_{ij}, \qquad \text{with} \qquad \eta_{eff} = \frac{\mu(I)P}{|\dot{\gamma}|}, \qquad (5)$$

and where $|\dot{\gamma}|$ is the second invariant of the shear rate tensor: $|\dot{\gamma}| = \sqrt{\frac{1}{2}\dot{\gamma}_{ij}\dot{\gamma}_{ij}}$.

Within this description, the granular liquid is described as an incompressible non-newtonian fluid, with an effective viscosity $\eta_{eff} = \mu(I)P/|\dot{\gamma}|$. This viscosity diverges when the shear rate $|\dot{\gamma}|$ goes to zero, which insures the existence of a flow threshold given by:

$$|\tau| > \mu_1 P \qquad \text{with} \qquad |\tau| = \sqrt{\frac{1}{2}\tau_{ij}\tau_{ij}}. \qquad (6)$$

This description is very similar to the one developed in other visco-plastic material like mud. However, granular matter is peculiar, because the viscosity depends on the pressure and not only on the shear rate.

In the following, we show that this approach predicts some important features of granular flows.

2.2. Example of application of the local rheology

2.2.1. Flow down an inclined plane.
Let us consider a granular layer flowing down a rough inclined plane (Figure 6a). We first consider the steady and uniform regime. The stress distribution in this configuration is given by $\sigma_{xz} = \rho g \sin\theta(h-z)$ and $\sigma_{zz} = P = -\rho g \cos\theta(h-z)$. One can then apply the constitutive law 3 to predict the velocity profile $u(z)$ and the volume fraction profile $\phi(z)$. The ratio between shear and normal stress being constant, one obtains the following relation:

$$\mu(I) = \tan\theta \quad \text{with} \quad I = \frac{d\,du/dz}{\sqrt{g\phi\cos\theta(h-z)}}. \tag{7}$$

This equation implies that the inertial number is constant across the layer, which directly implies that the volume fraction ϕ is independent of z. The velocity profile can also be integrated, assuming that the roughness condition corresponds to a zero velocity at the base. The predicted velocity varies like $z^{3/2}$ and is called a Bagnold profile [6, 101]:

$$\frac{u(z)}{\sqrt{gd}} = \frac{2}{3}I_0\frac{\tan\theta - \mu_1}{\mu_2 - \tan\theta}\sqrt{\phi\cos\theta}\left(\frac{h^{3/2} - (h-z)^{3/2}}{d^{3/2}}\right). \tag{8}$$

These predictions can be compared with experiments and numerical simulations. The first comparison concerns the flow threshold. According to Eq. (8) a steady and uniform flow is possible only if the inclination is comprised in between a minimum angle $\theta_1 = \arctan\mu_1$ and a maximum angle $\theta_2 = \arctan\mu_2$. This is observed in experiments, in which steady uniform flows are observed only in a range of inclination [80]. However, contrary to the prediction, the minimum angle to get a flow is not a constant, but depends on the thickness of the layer (Figure 6c). This critical angle $\theta_{\text{stop}}(h)$ is higher for thin layers than for thick layers, which is a signature of non-local effects. The second comparison concerns the shape of the velocity and volume fraction profile. For a thick layer, numerical simulations show that the flow is well described by the Bagnold profile (Figure 6b). The agreement is less good for thin layers, where the profile becomes more linear [88]. A constant volume fraction profile is also observed in the simulation. A last prediction concerns the evolution of the mean velocity. Experiments [80] and simulations [102] show that there exists a correlation between the depth averaged velocity $\bar{u}$, the thickness of the layer h and the inclination angle θ:

$$\frac{\bar{u}}{\sqrt{gh}} = \beta\frac{h}{h_{\text{stop}}}, \tag{9}$$

with $\beta \simeq 0.14$ is a constant and $h_{\text{stop}}(\theta) = \theta_{\text{stop}}^{-1}$ is the minimal thickness necessary to get a flow at inclination θ. The scaling of $\bar{u}$ with $h^{3/2}$ is compatible with the prediction of the local rheology $\mu(I)$. However, the angle dependence suggests a relation between the function $h_{\text{stop}}(\theta)$ and the friction law $\mu(I)$ [36]. Whether this link is a coincidence or reveals a more profound physical meaning remains an open question [27].

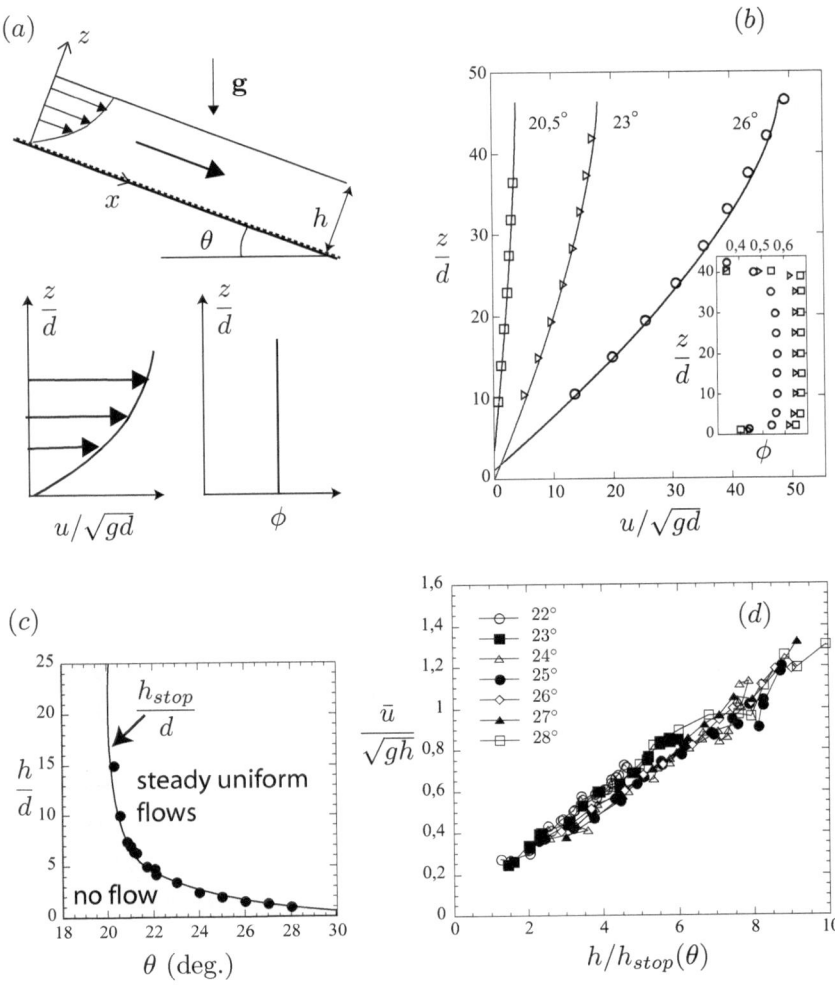

FIGURE 6. (a) Flow down inclined planes and the prediction of the local rheology; (b) Comparison between the Bagnold velocity profile (lines) and molecular simulation for spheres. Inset, volume fraction profile (data from [9]). (c) domain of existence of steady uniform flows. (d) Normalised depth averaged velocity as function of (h/h_{stop}) (from [80]).

The local rheology is then able to capture some characteristics of the steady uniform flows down inclined planes. One can go one step further and analyse the stability of such thin flows. It is well known with classical fluids that, when the flow becomes faster and faster, the free surface eventually becomes unstable and presents long wave modulations [93]. This instability is called the Kapitza instability for viscous liquids, roll waves instability for turbulent flows, and is also observed with mud [7]. It has been shown that the same kind of free surface deformation is also observed with granular materials (Figure 7 [32]). The granular roll waves have been experimentally investigated and the dispersion relation of the instability, namely how the small perturbations are amplified or attenuated depending on their frequency, has been measured using a well-controlled forcing method at the entrance of the flow [32]. These measurements provide a severe test for rheological models, as the characteristics of the instability strongly depend on the rheological properties of the liquid. In order to test the relevance of the local visco-plastic rheology, Forterre [33] has performed a linear stability analysis of the problem, using the tensorial formulation of the friction law Eq. (4). He has shown that once the parameters of the friction law are calibrated using the steady uniform flows, the theory gives quantitative predictions for the instability threshold and the dispersion relation of the instability (Figure 7). This study shows that the proposed local rheology is relevant to describe non-trivial three-dimensional flows.

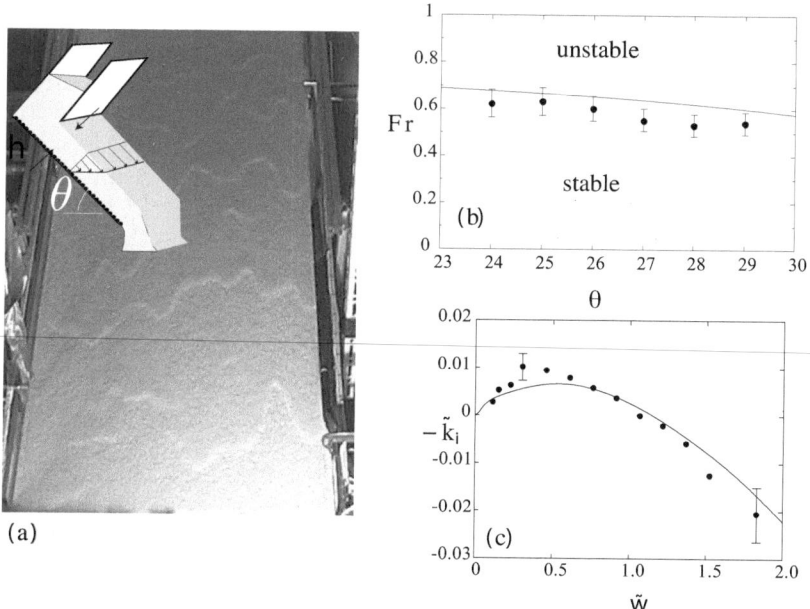

FIGURE 7. (a) picture showing the free surface waves observed when a granular media flows down an inclined plane. (b) Instability threshold in terms of the Froude number versus the inclination. lines: prediction of the local rheology, markers, experimental measurements. (c) spatial growth rate versus frequency (from [33]).

88   Y. Forterre and O. Pouliquen

2.2.2. Flow on a pile. Another interesting configuration to study the granular rheology is the flow on a pile, obtained when a granular layer flows on a static heap (Figure 8). Contrary to the case of the flow down an inclined plane, the inclination θ and the thickness h of the flow are not imposed by the experimentalist, but are selected by the system itself. The only control parameter is the flow rate Q. In this configuration, it has been shown that steady and uniform flows are possible if the system is confined in between two walls. In this case, the additional friction induced by the lateral walls plays a crucial role and is responsible for the localisation of the flow at the free surface [104, 50]. One can try to model this configuration using the visco-plastic description, by taking into account the lateral boundary conditions. In the case of lateral rough walls, a typical velocity profile predicted by the theory is plotted in Figure 8a. The model correctly predicts the shape of the profile with a localisation close to the free surface. The agreement is not only qualitative, but quantitative predictions can be made for the free surface velocity, once the friction law $\mu(I)$ is calibrated as explained in the previous section (Figure 8b) [51]. This is a second example of a flow with a shear deformation in different directions, which is well described within the framework of a local rheology. However, the description is not perfect and some experimental observations are not well captured by the model. First a transition from a continuous flow regime to an avalanching regime is observed when the flow rate decreases [62, 51]. This transition is not predicted by the model. Secondly, the interface between the flowing region and the static pile is not as discontinuous as predicted by the theory. Experimentally, a slow creep is observed in the static region, with an exponential tail which is not predicted by the local rheology [56].

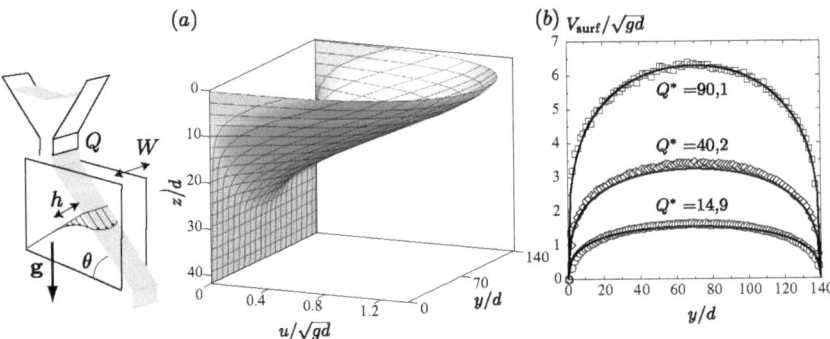

FIGURE 8. Flow on pile. (a) 3D velocity profile predicted by the visco-plastic local rheology. (b) quantative comparison between the theory and experiments for the velocity profile observed at the free surface (from [51]).

2.2.3. Granular collapse. A flow configuration, which has attracted a lot of attention the last ten years, is the collapse of a granular column under gravity. A cylinder full of grains is suddenly lifted up. The material then spreads over the surface. This configuration is a model for cliff collapses in geophysics [59, 68, 8]. Experiments have revealed interesting scaling for the spreading distance as a function of the aspect ratio of the initial column. However a complete description is still lacking. An interesting question is whether the local visco-plastic approach would be able to correctly predict the dynamics of this fully three-dimensional flow. However, implementing the rheology in three-dimensional fluid mechanics code is a tricky work [35], which to our knowledge is not yet completely achieved. However, a recent study performed by Lacaze and Kerswell [58] suggests that the approach may be relevant to describe the whole dynamics. This authors have performed numerical simulations of the collapse problem using molecular dynamics simulations. Knowing at each time, the position of the particles, their velocities, the forces at each contact, their were able by a suitable coarse graining process to compute the shear rate, the shear stress, the pressure, and check at each position, at each time, how the friction coefficient varies with the inertial number. Figure 9 shows that all the points collapse quite well along a line, which has the same shape as the one obtained in simple configurations as plane shear. This result gives good hope that the visco-plastic description is enough to capture most of the dynamics of this complex three-dimensional flows.

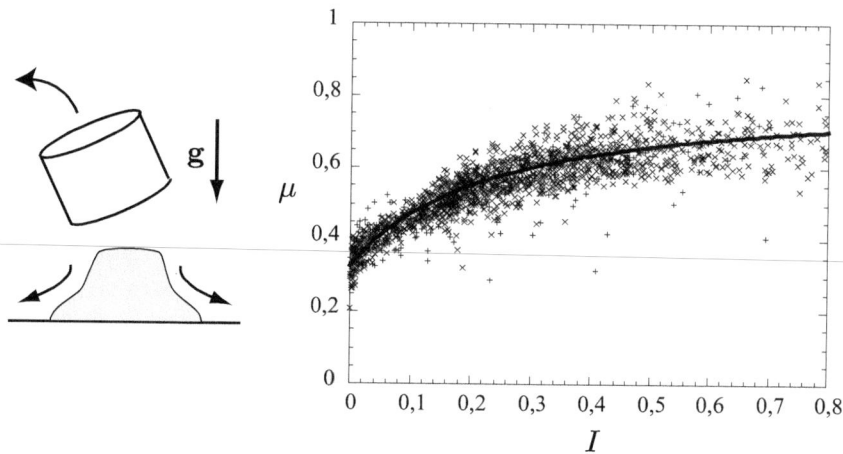

FIGURE 9. Collapse of a granular column under gravity simulated using molecular dynamics. Each point corresponds to the ratio $\mu = \tau/P$ as a function of I for different positions, different times, different aspect ratios. The continuous curve is the best fit (from [58]).

2.2.4. Confined flows. The flows considered in the previous sections are free surface flows. Other geometries, in which the material is confined in between walls, have been also intensively studied, including the cylindrical Couette cell [106](Figure

3b), the vertical silo (Figure 3c), and the plane shear with gravity ([10, 36] and references therein). In all these configurations, the velocity profile is localized in a shear band 5 to 10 particles thick located close to the moving wall. In more complex 3D geometries such as the modified Couette cell, where the bottom is split in a rotating and a static part, shear zones up to 40 particle diameters wide are observed [30]. It is important to keep in mind that all these flows are most of the time conducted in a quasi-static regime, for which the inertial number is less than 10^{-4}. In this regime, the $\mu(I)$ rheology reduces to a simple frictional Drucker-Prager plasticity criterion (Eq. (6)) and does not correctly predict the shear bands observed experimentally. The localization of the shear close to the moving wall is predicted, as it is due to a non-uniform stress distribution. However, the thickness of the predicted shear bands depends on the shear velocity and vanishes in the quasi-static limit. This is in contradiction with the observations and clearly shows that the local rheology is not able to capture the quasi-static regime.

Although the $\mu(I)$ rheology does not correctly predict the width of the shear bands, it can be useful to predict their position in cases where a complex 3D pattern develops. For example, in the case of the flow induced by the rotation of a disk in a granular media, the shear band takes the form of a cap (Figure 10), or a column. The local rheology captures the correct shape and the transition between a cap and a column state, depending on the aspect ratio. In this example the viscous part of the visco-plastic rheology does not play any role, but allows us to approach this quasi-static problem from a fluid mechanics point of view, which appears to be easier than from a pure plasticity point of view.

2.3. Beyond $\mu(I)$

The previous examples have shown that the phenomenological rheology $\mu(I)$ allows us to describe many properties of dense granular flows. However, we have seen that some characteristics are not predicted by this simple approach. In this section we discuss the different limits of the local approach and the other theoretical attempts developed to describe granular flows.

2.3.1. The solid-liquid transition: Role of the preparation, hysteresis, and finite size effects.

The first limit of the $\mu(I)$ rheology concerns the starting and stopping properties. Within the model, the flow threshold is described by a unique friction angle μ_1, which corresponds to a simple Coulomb criterion. However, the transition between flow and no flow in a granular medium is a more complex phenomenon.

First, the way the material starts depends on the initial preparation of the sample. Both the initial volume fraction and the history of the previous deformation play a role [21]. In order to describe these effects, it is necessary to introduce additional internal variables, like the volume fraction, the texture, which characterize the anisotropy of the force network. Attempts exist within plasticity models [92], but the link with the $\mu(I)$ rheology remains to be done.

A second limit of a simple Coulomb criterion is that it does not describe the hysteresis observed in some flow configurations. For example, let us consider the

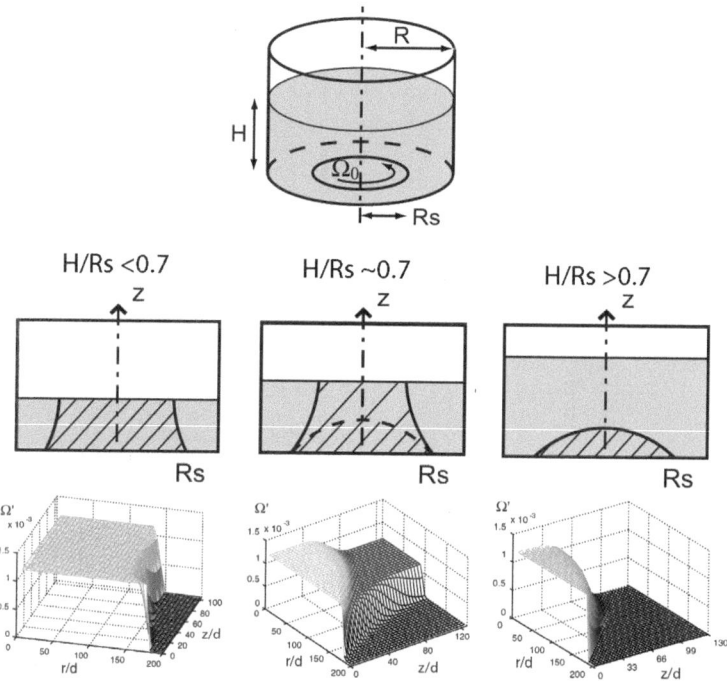

FIGURE 10. Shear bands created by a split bottom container. Depending on the aspect ratio, the granular media entrained by the disk take different forms, well predicted by the simulation of the local rheology (bottom line) (figure from [52]).

case of a granular layer on a rough inclined plane. Starting from a static layer of thickness h, one has to incline the plane up to a critical angle θ_{start} in order for the flow to start. Once the layer moves, one has to decrease the inclination below θ_{stop} less than θ_{start} in order to stop the flow [79]. In between these two angles, the system is metastable: a small perturbation can be enough to trigger an avalanche. Depending on the inclination, the avalanche can propagate down the slope only or can go uphill and put into motion the whole layer [20] (Figure 11a). This kind of hysteresis is also observed in other configurations when the system is driven by the stress, for example in a Couette cell when imposing a torque on the inner cylinder [36]. The physical origin of the hysteresis is not clear, although an analysis based on the dynamics of a single grain on a bumpy surface shows that it is related to the balance between the external stress, the dissipation due to collision and the geometrical traps formed by the bump [86, 2]. A phenomenological theoretical approach has been developed [107, 4] to describe this hysteresis. The granular media is described as a mixture of solid and liquid, which proportion is controlled by an order parameter.

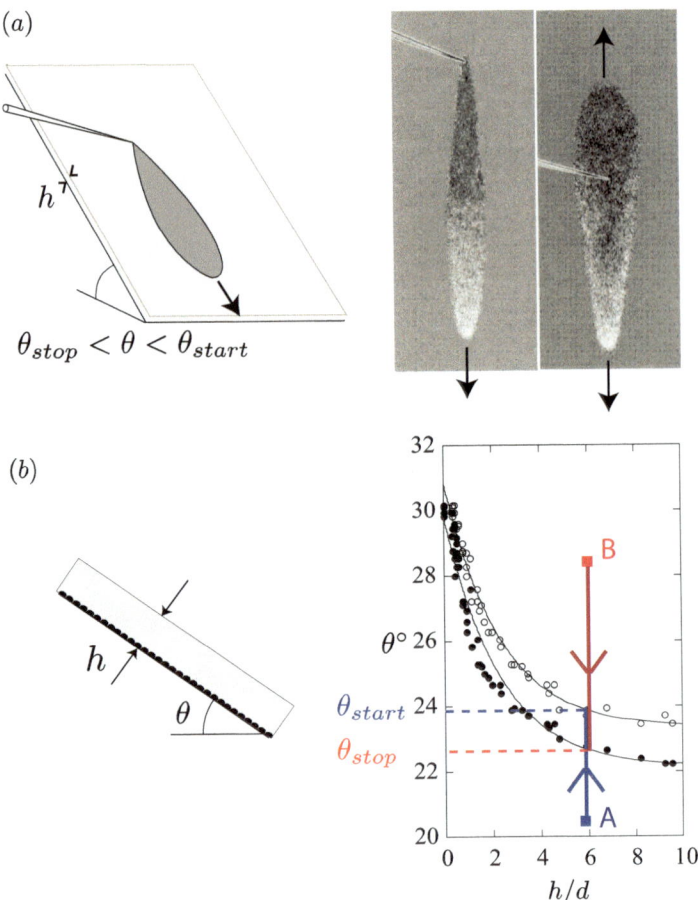

FIGURE 11. (a) Avalanches on a thin layer of grains initially static on a rough plane inclined in a metastable state. Depending on the inclination angle, the avalanche, which is triggered by a tiny perturbation, propagates downhill only or uphill (from [20]). (b) starting and stopping angles of a granular layer on a rough incline for glass beads (from [31]).

    The last limit of the simple Coulomb description of the solid-liquid transition of a granular medium concerns finite size effects. In the case of the inclined plane for example, the starting angle θ_{start} and stopping angle θ_{stop} depend on the thickness of the layer h, as shown in Figure 11b [79, 80]. For thick layers larger than 20 particle diameters, these angles are independent of h. However, for thin layers, both angles increase. This additional rigidity of a thin layer compared to a thick layer is not yet understood, but could be related to non-trivial collective effects.

2.3.2. Quasi-static flows. The second serious limit of the local rheology $\mu(I)$ concerns the description of quasi-static flows. We have seen that for confined flows in a quasi-static regime the rheology correctly captures the location of the shear bands but fails in predicting their thickness, which goes to zero when I goes to zero, in contradiction with the observations. For flow on a heap also, far from the free surface, the rheology predicts a zero velocity with a true solid part, whereas experimentally an exponential tail is measured, corresponding to a slow creep motion. These observations of a slow creep motion on typically 10 particle diameters suggests that the simple assumption of a local rheology, i.e., a one-to-one relation between the shear rate and the stresses, is wrong. Several approaches have been proposed to describe these quasi-static flows. The first consists in modifying plasticity models to take into account fluctuations of stresses [53], or rotation [73]. A second approach consists in writing non-local rheological laws [72, 85]. One idea which seems to emerge from the different attempts to describe this regime is the role played by the mechanical noise and correlation [11, 84, 87]. In all the athermal systems such as foams, glasses and granular systems, a rearrangement somewhere induces stress and strain fluctuations, which can in turn influence deformation somewhere else. How to take into account such non-local effects in constitutive equations remains an open question.

2.3.3. Transition liquid/gas and link with the kinetic theory. A last limit of the local rheology concerns the transition towards the gaseous regime described by the kinetic theory of granular gases [14, 38, 46, 47, 69]. This transition is much less studied than the solid-liquid transition, although it can be observed in several configurations [64, 65]. For example, it can be observed when a flow layers on a very steep plane. It does not reach a steady regime but accelerates and becomes more and more dilute [31]. In a flow on a heap confined between two walls also, if the flow rate is too large, a gaseous layer develops at the free surface [67, 50]. This transition to a very dilute regime is not predicted by the simple $\mu(I)$ approach, but it is well described by the kinetic theory of granular flows. However, the kinetic theory does not predict the correct behaviour in the dense flow regime, and predicts a friction coefficient μ which decreases with the inertial number [34]. Trying to reconcile the two approaches has motivated several theoretical works trying to modify the standard kinetic theory of granular gases to be compatible with dense flow regimes [49, 66, 48, 57].

2.3.4. Conclusion about the local rheology. We have shown that, to the first order, the behaviours of a dense granular flow can be described using simple dimensional arguments and the assumption of a local rheology. Within this approach, the granular medium is described as frictional visco-plastic liquid, with a friction coefficient depending on the shear rate. This phenomenological approach predicts with success many flow configurations. However, serious limits exist concerning the transition to the static regime or the rapid regime. Morevover, most of the studies deal with spherical particles and the generalisation to more complex material media is another open question.

We show in the next section that the lack of more precise information about the rheology can in some cases be circumvented by writing depth-averaged conservation equations. By depth averaging, it is no longer necessary to specify a bulk rheology of the material, an expression for the basal stress being sufficient. This depth-averaged approach is very useful in many geophysical contexts.

3. Depth-averaged approach

Depth-averaged or Saint-Venant equations were introduced in the context of granular flows by Savage & Hutter [98]. The initial motivation was to model natural hazards such as landslides or debris flows [75, 42]. Assuming that the flow is incompressible and that the spatial variation of the flow takes place on a scale larger than the flow thickness, one obtains the Saint-Venant equations by integrating the three-dimensional mass and momentum conservation equations. For two-dimensional flows down a slope making an angle θ with the horizontal (see Figure 12), the depth-averaged equations reduce to

$$\frac{\partial h}{\partial t} + \frac{\partial h \langle u \rangle}{\partial x} = 0, \tag{10}$$

$$\rho_s \phi \left(\frac{\partial h \langle u \rangle}{\partial t} + \alpha \frac{\partial h \langle u \rangle^2}{\partial x} \right) = \left(\tan \theta - \mu_b - K \frac{\partial h}{\partial x} \right) \rho_s \phi g h \cos \theta, \tag{11}$$

where h is the local flow thickness, $\langle u \rangle = Q/h$ is the depth-averaged velocity (Q being the flow rate per unit of width), and ϕ is the volume fraction, assumed constant.

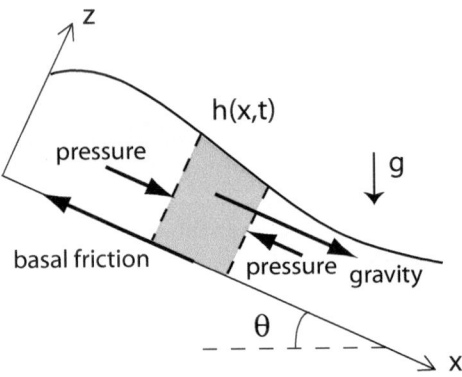

FIGURE 12. Forces balance in shallow water description.

Equation (10) is the mass conservation and (11) is the momentum equation, where the acceleration is balanced by three forces (Figure 12): the gravity parallel to the plane, the tangential stress between the fixed bottom and the flowing layer (written as a basal friction coefficient μ_b times the normal stress), and a pressure force related to the thickness gradient. The coefficient α is related to the assumed velocity profile across the layer and is of order 1. The coefficient K represents the ratio of the normal horizontal stress (x-direction) to the normal vertical stress (z-direction) and is close to 1 for steady uniform flows [101]. The main advantage of the Saint-Venant equations is that the dynamics of the flowing layer can be predicted without knowing in detail the internal structure of the flow. The complex three-dimensional rheology of the material is mainly embedded in the basal friction term μ_b.

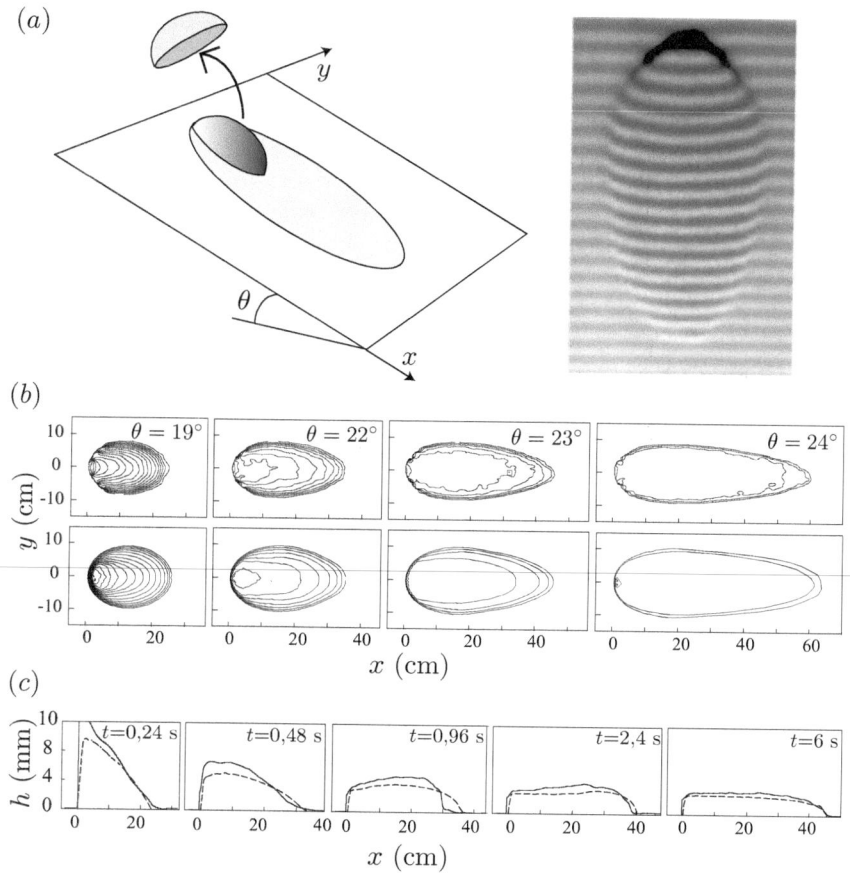

FIGURE 13. (a) spreading of a granular mass down an inclined plane measured by a Moiré (picture). (b)Final deposit as a function of the inclination. Comparison between experiment (top row) and depth averaged equations (bottom row). (c) Spreading dynamics as a function of time: experiment (solid line) and theory (broken line).

Taking a simple constant Coulomb-like basal friction is sometimes sufficient to capture the main flow characteristics [98] and has been used to describe granular slumping [8, 60, 70], rapid flows down smooth inclines [41, 108], and shock waves [40, 43]. However, for flows down rough inclines, the assumption of a constant solid friction is not compatible with the observation of steady uniform flows over a range of inclination angles. One can then use the local rheology developed in the previous section to propose an expression for the basal friction. In order to properly capture the hysteresis and the influence of finite size, more complex basal friction laws $\mu_b(\langle u \rangle, h)$ have been proposed, which lead to quantitative predictions in complex situations such as a propagating steady front, mass spreading, or surface instabilities [81, 82, 32, 71] (Figure 13). It should be noticed that the Saint-Venant Equations (11) and (12) represent a first-order development in terms of the flow aspect ratio. Therefore, they do not capture second-order effects like longitudinal and lateral momentum diffusion, which stabilize instabilities [33] and control lateral stresses. The knowledge of the full 3D constitutive equations (Eq. (4)) may allow the development of more complex depth-averaged models [93, 7].

Another application of the depth-averaged equations concerns situations where the flowing layer propagates on an erodible layer, such as flows on top of a static pile. In this case, an exchange of matter exists between the liquid and solid phase. An additional equation is then needed to determine the solid-liquid interface. Several closures have been proposed [3]. The first model [12, 13] assumes that erosion/deposition is controlled by the difference between the local slope and the critical pile angle. Other approaches assume a relation between the averaged velocity and the flow thickness, either by fixing the velocity gradient [25] or by prescribing a basal shear stress at the solid/liquid boundary [54]. These models predict qualitatively non-trivial behaviours such as the propagation of avalanche fronts [26, 105]. Although these two-layer approaches seem a promising framework to study avalanching flows on erodible beds, it is important to note that the closures proposed to date are not compatible with observations of steady uniform flow on a pile, where the flowing thickness is selected by the side walls [50]. This clearly shows that a proper development of shallow water models has to rely on the knowledge of the full constitutive equations, a goal not yet completely achieved.

To conclude this section on depth-averaged equations, it is important to emphasize that this framework is often used to model real situations encountered in geophysics. It is possible to take into account a complex topography by considering an inclination which varies with space and by adding additional centrifugal forces [39]. Examples of simulation of rock avalanches and pyroclastic flows are shown in Figure 14.

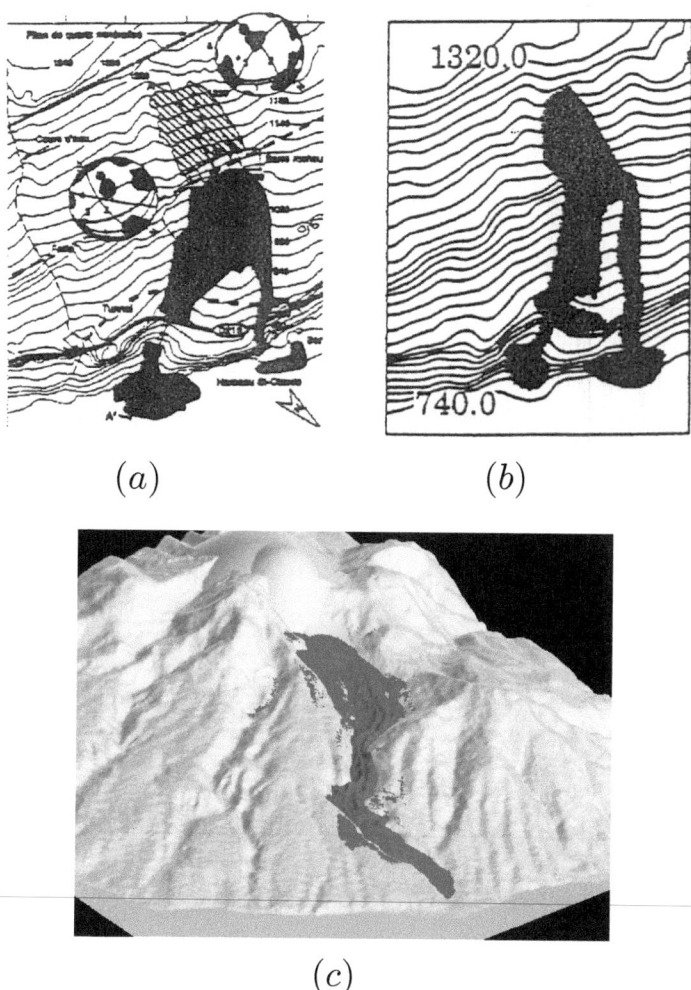

(a)   (b)

(c)

FIGURE 14. Applications of depth-averaged equations to natural events
(a) trajectory observed for the land slide of Charmonetiers, Isere 1987
and simulation (b) from [75]. (b)simulation of the boxing day event
(December 26th 1997), Montserrat Island, Lesser antilles (from [42]).

4. Towards more complex granular materials

We have seen that, in the case of a simple granular medium, some advances have
been made in the description of the flow properties. In this section we discuss
to what extent this progress in our understanding of granular flows can help in
describing more complex material encountered in applications. We discuss the

role of the polydispersity, the role of cohesion and the role of the presence of an interstitial liquid.

4.1. Polydispersed media

Most of the granular media encountered in applications are made of grains having different sizes. A major problem, which arises when manipulating polydispersed material, is the size segregation [76, 97, 29]. During the flow, particles of different sizes tend to separate. Despite its importance in industrial processes, and despite the large number of studies, the mechanisms responsible for segregation are far from being understood. Different scenarii have been proposed (percolation, statistic sieving [97]). A review of the research on segregatation is far beyond the scope of this paper. In this section, we will focus on the rheology of polydispersed material and discuss how the friction law evidenced in monodispersed granular media can be modified to account for the presence of different grains.

For monodispersed material the rheology is given by a friction coefficient depending on the inertial number $I = \dot{\gamma}d/\sqrt{P/\rho_p}$, where d is the particle diameter. It is then tempting to generalize this approach to the case of polydispersed material by using the local mean particle diameter $\bar{d}$ in the definition of I:

$$I_{\bar{d}} = \frac{\dot{\gamma}\bar{d}}{\sqrt{P/\rho_p}}. \tag{12}$$

This idea has been recently tested for the flow of disks down inclined planes [90]. In this study, the material is composed of two different sizes: large disks of diameter d_b and small ones of diameter d_s. As for monodispersed material, there exists a range of inclination for which a steady uniform flow develops. Segregation induces a non-uniform distribution of large particles across the layer: the large are at the free surface, whereas the small concentrate at the bottom. The authors introduced a local mean diameter defined by:

$$\bar{d}(z) = \frac{\phi_s(s)d_s + \phi_b(z)d_b}{\phi(z)}, \tag{13}$$

where ϕ is the volume fraction of the medium, and ϕ_s and ϕ_b are the local volume fraction of small and large particles respectively ($\phi = \phi_s + \phi_b$). The first results obtained by Rognon *et al.* [90] is that the inertial number computed using this mean diameter is constant across the layer in agreement with the local rheology. Consequently, if $I_{\bar{d}}$ is constant, this implies that $\dot{\gamma} \propto 1/\bar{d}$ for a given inclination and position. The velocity gradient is then inversely proportional to the mean size of the particles. This prediction is observed in the simulation: the shear rate is higher at the bottom where there is an accumulation of small particles, and decreases at the top where large particles migrate. This result strongly suggests that the frictional visco-plastic rheology may be relevant for polydispersed material. However, if the rheology tells us how the change in relative concentration of species changes the flows, it does not predict the segregation leading to the distribution of grains. This point remains a challenge in the physics of granular media.

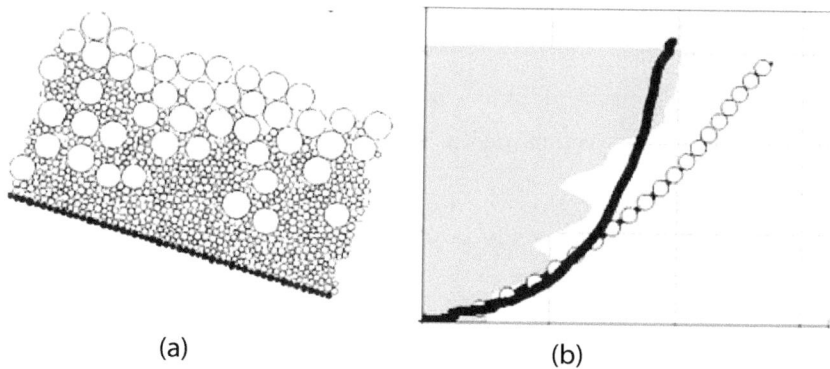

(a)   (b)

FIGURE 15. Flow of bidispersed material on an inclined plane in 2D (a)
Sketch of the flow. (b) Velocity profile (solid line) compare to velocity
profile observed in the monodispersed case (symbols). The shaded region
corresponds to the concentration of large particles (from [90]).

4.2. Cohesive granular media

The local rheology can also be generalized to the case of cohesive material. In
this case one can consider that the grains interact not only by contact interaction
but that an additional attractive force exists, which tends to put the grains in
contact. The origin of cohesion can be the Van der Waals interactions, capillary
bridges, electrostatic forces ... Taking into account in the rheology the details of
these cohesive interactions is still an open question. However, some authors have
studied a simpler case [89], in which the force interaction is simply characterized
by a maximum force N_c. The interaction force is then zero if the particles are not
in contact, decreases to a maximum $-N_c$ and increases again when approaching
the grains further. The force then becomes again positive, i.e., repulsive, when
the overlap between particles becomes large (one recovers the repulsive elastic
interaction) (Figure 16). This model represents the simplest cohesive material and
allows the authors to capture the major features of the flow of cohesive granular
media.

To study the rheology of this simple cohesive material, one can consider the
plane shear configuration (Figure 4) where the material is sheared at a constant
shear rate $\dot{\gamma}$ and confined under a pressure P. This analysis has been carried
out by Rognon et al. [89] using molecular dynamics simulation. In this case also,
the dimensional analysis of the problem appears to be very fruitful. By contrast
with the dry case for which the inertial number $I = \dot{\gamma}d/\sqrt{P/\rho_p}$ was the single
dimensionless control parameter of the problem, the case of the cohesive material
introduces a second dimensionless number C, which is the ratio of the maximum
attractive force N_c divided by the characteristic pressure force Pd^2 (Pd in 2D):

$$C = \frac{N_c}{Pd^2}. \qquad (14)$$

One can then directly conclude that the friction coefficient and the volume fraction can be written as follows:

$$\mu = \mu(I, C) \quad \text{and} \quad \phi = \phi(I, C). \tag{15}$$

Rognon *et al.* [89] have systematically studied the variation of μ and ϕ with I and C (Figure 16).

Although this study concerns an oversimplified model of cohesive material, it suggests that the progress made in our understanding of the rheology of dry granular materials can serve as a base to develop rheological models for cohesive material.

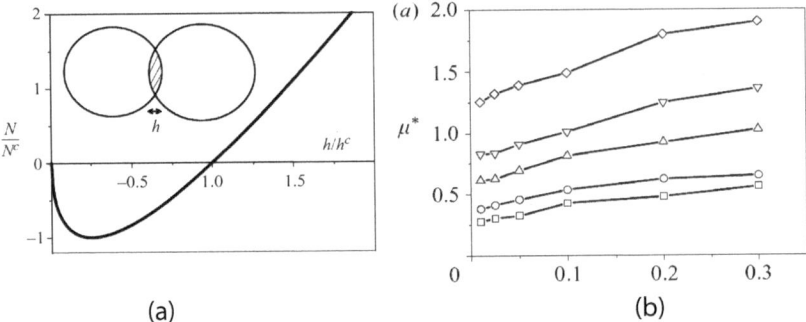

(a)   (b)

FIGURE 16. (a) Interaction force versus overlap between particles for the model of cohesive material studied in [89].(b) Friction coefficient as a function of the inertial number for different values of the cohesion parameter C ($C = 0, 10, 30, 50, 70$ from top to bottom) (from [89]).

4.3. Immersed granular media

The last example we would like to discuss concerns the case of granular media immersed in a liquid. To what extent does the presence of the liquid modify the rheology of the material, and is it possible to propose constitutive equations for the granular paste? A simple way to address this question consists again in considering the plane shear configuration under constant pressure. The grains immersed in a liquid are confined by a porous plate which applies a pressure P^p on the particle and imposes a shear rate $\dot{\gamma}$. As in the dry case, we want to know how the shear stress τ^p and the volume fraction varies with $\dot{\gamma}$ and P^p. We have seen that the important idea is to compare the typical time of deformation $t_{\text{macro}} = 1/\dot{\gamma}$ and the typical time of rearrangement t_{micro}. If the deformation is slow compared to the typical time it takes for a particle to fall in a hole, it can be considered as a quasi-static deformation. This picture suggests a first naive approach of the rheology of immersed granular media. The presence of the fluid is going to change the typical falling time of a grains t_{micro} and will then change the constitutive law of the material.

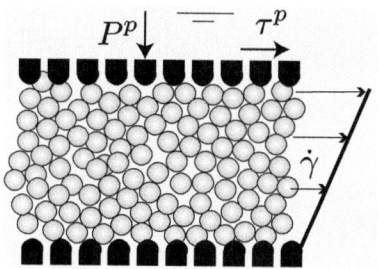

FIGURE 17. The plane shear configuration under constant normal stress for an immersed granular material.

The study of the time taken by a particle to fall in a fluid under a pressure P_p has been done in [16, 15] and has put in evidence different regimes. The equation controlling the motion of the particle of mass m can be written as follows:

$$m\frac{d^2z}{dt^2} \simeq P^p d^2 - F_{\text{drag}}. \tag{16}$$

One can then distinguish between three regimes:

- *Free fall regime.* In this regime the drag induced by the fluid is negligible during the fall. The particle during its motion follows an accelerated motion described by the two first terms of Eq. (16). This is the dry regime discussed before. Considering that $z \simeq d$ and $t \simeq t_{\text{micro}}$ in the first term, we get $t_{\text{micro}}^{\text{fall}} = d/\sqrt{P^p/\rho_p}$.
- *Viscous regime.* In this regime the grain rapidly reaches its terminal velocity given by the balance between the viscous drag and the pressure. Knowing that $F_{\text{drag}} \simeq \eta d \, dz/dt$ one finds that $t_{\text{micro}}^{\text{visc}} \simeq \eta/P^p$.
- *Inertial regime.* In this regime the grain also reaches its terminal velocity but it is controlled by the inertial drag force given by $F_{\text{drag}} \simeq C_d d^2 \rho_f (dz/dt)^2$ where C_d is the drag coefficient. One gets that $t_{\text{micro}}^{\text{inert}} \simeq d/\sqrt{P^p/(\rho_f C_d)}$.

The transition between the different regimes is then controlled by two dimensionless numbers: a Stokes number St, which is the ratio between the free fall time scale over the viscous time scale, and the number r ratio of the free fall time over the inertial time scale

$$St = \frac{t_{\text{micro}}^{\text{fall}}}{t_{micro}^{\text{visc}}} \simeq \frac{d\sqrt{\rho_p P^p}}{\eta}, \tag{17}$$

$$r = \frac{t_{\text{micro}}^{\text{fall}}}{t_{\text{micro}}^{\text{inert}}} \simeq \frac{\rho_p}{\rho_f C_d}. \tag{18}$$

A phase diagram can then be drawn in the parameters space (St, r). If the longest time is $t_{\text{micro}}^{\text{fall}}$ (if $St \gg 1$ and $r \gg 1$), one gets the regime of dry granular flows, for which the fluid is negligible. If the longest time is $t_{\text{micro}}^{\text{visc}}$, (if $St \ll 1$ and

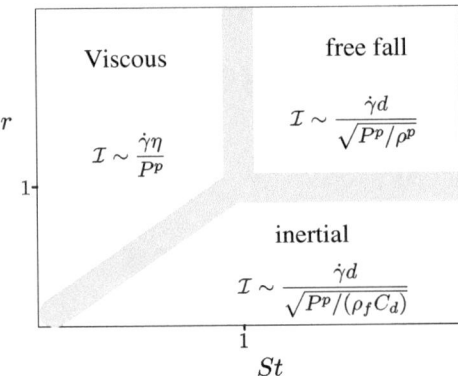

FIGURE 18. The different flow regimes in the plane (St,r) for immersed granular flows sheared under a confining pressure P^p. The expression of the dimensionless number $\mathcal{I}$ is given in each regime.

$r \ll St$) the regime is a viscous regime. Finally, if the longest time is $t_{\text{micro}}^{\text{inert}}$ (if $St \ll r$ and $r \ll 1$), the regime is inertial.

When coming back to the problem of the plane shear configuration, one can then suggest that the relevant dimensionless number is going to be given by the shear rate multiplied by the microscopic time scale. We then call this number $\mathcal{I}$. The different expressions of $\mathcal{I}$ for the three regimes are presented in Figure 18.

From this analysis one can then propose that the constitutive laws describing the rheology of immersed granular material are given by the friction law and the volume fraction law:

$$\tau^p = \mu(\mathcal{I})P^p \quad \text{and} \quad \phi = \phi(\mathcal{I}). \tag{19}$$

One recovers the rheology of dry granular material in the free fall regime, but other scaling laws are obtained in the other regimes. To our knowledge, there is no precise test of such a phase diagram based on the analysis of the typical time scales, and no direct measurement of the function $\mu(\mathcal{I})$ and $\phi(\mathcal{I})$ are available (except in the dry regime as discussed previously).

However, in the viscous regime, indirect measurements exist derived from experiments of flows down an inclined plane. These measurements show that the friction law as a function of the viscous number $\mathcal{I}$ follows a shape similar to the one observed in the dry case [15, 78]. This approach has been successfully applied to describe the immersed flow down a pile confined between lateral walls [24].

Up to now, we have discussed the rheology of immersed granular media sheared under a constant pressure, by analogy with the dry granular case. This configuration is relevant for free surface flows like submarine avalanches, in which the gravity prescribes the stress. However, this situation is not conventional in the field of the rheology of suspensions. Most of the studies concern shear at constant volume fraction, in a situation where particles have the same density as the fluid

[103]. This is equivalent to keeping the top plate at a fixed distance in Figure 17. In this case, the pressure P_p on the top plate is no longer a control parameter, but has to be measured. In this configuration, the existence of different regimes has been discussed in several papers ([1, 61]). The viscous regime has attracted most of the studies on the rheology of suspensions. Under the assumption that inertia does not play any role, dimensional analysis implies that the shear stress τ_p and the normal stress P_p measured at the top plate varies with $\eta\dot{\gamma}$ [74, 61]:

$$\tau_p = f_1(\phi)\eta\dot{\gamma} \quad \text{et} \quad P_p = f_2(\phi)\eta\dot{\gamma}. \tag{20}$$

The function $f_1(\phi)$ is the relative viscosity, which has been measured in a wide range of ϕ. $f_1(\phi)$ increases with ϕ and seems to diverge at a maximum volume fraction [44]. The normal stress $f_2(\phi)$ has been much less studied. Discussing the different models and implications of such rheology for dense suspensions is far beyond the scope of this paper dedicated to granular flows. The only important point we would like to underline, is the analogy that exists between the granular approach (19) and the suspension approach (20). The two expressions are the same if one chooses $\phi(I) = f_2^{-1}(1/\mathcal{I})$ and $\mu(\mathcal{I}) = \mathcal{I}f_1(f_2^{-1}(1/\mathcal{I}))$. For the two expressions to be really identical, the divergence of $f_2(\phi)$ when approaching the maximum volume fraction has to be the same as the divergence of $f_1(\phi)$, in order to have a friction coefficient τ^p/P^p going to a constant. This analogy suggests that immersed granular media and suspension could be described within the same framework.

In the simple shear cell configuration we have discussed, there is no relative motion between the fluid and the granular material. The fluid moves with the grains. However, there are configurations for which it is no longer the case. One example is the sediment transport. When a liquid flows on top of a granular bed, the fluid puts the grains into motion. In this case, the drag force induced by the velocity difference between the fluid and the grains is the motor of the granular flows. Another example in which the fluid moves relatively to the grains is the initiation of avalanches. When a granular matter starts to deform, it compacts or dilates depending on the initial preparation. In presence of a liquid, a dilatation (resp. a compaction) of the granular layer means that the liquid has to flow into (resp. out of) the granular layer. This fluid motion induces an additional stress on the granular skeleton, which modifies the deformation. A framework, which seems to be relevant to address this coupling problem, is the two-phase flow equations framework. The idea is to consider the liquid phase and the solid phase as two continuum media, and in writing the mass and momentum conservation equations for the two phases. The challenge lies in proposing an expression for the stresses in each phase. The simple idea we have developed in this section based on the friction law and the volume fraction law (Eq. (19)) gives a relevant suggestion for the rheology of the granular phase. Some recent works on sediment transport [77] and on the triggering of submarine avalanches [78] seem to show the relevance of this approach.

5. Conclusion

We have presented a survey of our current understanding of dense granular flows. Our main intention was to emphasize that a zero-order description of the viscous-like behavior of dense granular flows is now available, which relies on simple but solid dimensional arguments. A frictional visco-plastic formulation has been developed which gives quantitative predictions for different flow configurations and can serve as a first tool to predict other configurations encountered in applications. Although promising, this approach fails to capture the details of the quasi-static flows and the transition to solid or gaseous regimes. It is difficult to anticipate that more elaborated constitutive equations will be developed in the near future that can describe the whole phenomenology of granular flows. The diversity of the theoretical approaches clearly shows that the task is difficult, the central question being, in our opinion, how to take into account non-local effects created by the network of enduring contacts.

We have also discussed in this paper how recent progress in our understanding of simple dry granular flows can serve as a base to tackle more complex materials such as the one used in industry or encountered in geophysics. The role of cohesion, the role of polydispersity, the role of the interstitial fluid has been discussed. Fundamental research on such complex granular materials is very young, and no doubt that the intense activity will give rise in the next future to important progress.

References

[1] C. Ancey, P. Coussot, P. Evesque, *A theoretical framework for granular suspension in a steady simple shear flow*, J. Rheology **43** (1999), 1673–99.

[2] B. Andreotti, *A mean field model for the rheology and the dynamical phase transitions in the flow of granular matter*, Europhys. Lett. **70** (2007), 34001.

[3] A. Aradian, E. Raphaël, P.G. de Gennes, *Surface flows of granular materials: a short introduction to some recent models*, C. R. Physique **3** (2002), 187–96.

[4] I.S. Aranson, L.S. Tsimring, *Continuum theory of partially fluidized granular flows*, Phys. Rev. Lett. **65** (2002), 061303.

[5] I.S. Aranson, L.S. Tsimring, *Patterns and collective behavior in granular media: theoretical concepts*, Rev. Mod. Phys. **78** (2006), 641–692.

[6] R.A. Bagnold, *Experiments on a gravity-free dispersion of large solid spheres in a Newtonian fluid under shear*, Proc. R. Soc. Lond. A **225** (1954), 49–63.

[7] N.J. Balmforth, J.J. Liu, 2004. *Roll waves in mud*, J. Fluid Mech. **519** (2004), 33–54.

[8] N.J. Balmforth, R.R. Kerswell, *Granular collapse in two dimensions*, J. Fluid Mech. **538** (2005), 399–28.

[9] O. Baran, D. Ertas, T.C. Halsey, G.S. Grest, J.B. Lechman, *Velocity correlations in dense gravity-driven granular chute flow*, Phys. Rev. E **74** (2006), 051302.

[10] L. Bocquet, W. Losert, D. Schalk, T.C. Lubensky, J.P. Gollub, *Granular shear flow dynamics and forces: experiments and continuum theory*, Phys. Rev. E **65** (2001), 011307.

[11] D. Bonamy, F. Daviaud, L. Laurent, M. Bonetti, J.P. Bouchaud, *Multi-scale clustering in granular surface flows*, Phys. Rev. Lett. **89** (2002), 034301.

[12] J.P. Bouchaud, M. Cates, J.R. Prakash, S.F. Edwards, *A model for the dynamics of sandpile surfaces*, J. Phys. France I **4** (1994), 1383–1410.

[13] T. Boutreux, E. Raphaël, P.G. de Gennes, *Surface flows of granular material: a modified picture for thick avalanches*, Phys. Rev. E **58** (1998), 4692–4700.

[14] C.S. Campbell, *Rapid granular flows*, Ann. Rev. Fluid Mech. **22** (1990), 57–92.

[15] C. Cassar, M. Nicolas, O. Pouliquen, *Submarine granular flows down inclined planes*, Phys. Fluids **17** (2005), 103301.

[16] S. Courrech du Pont, P. Gondret, B. Perrin, M. Rabaud, *Wall effects on granular heap stability*, Europh. Let. **61** (2003), 492–98.

[17] F. Da Cruz, F. Chevoir, D. Bonn, P. Coussot, *Viscosity bifurcation in granular materials, foams, and emulsions*, Phys. Rev. E **66** (2002), 051305.

[18] F. Da Cruz, F. Chevoir, J.N. Roux, I. Iordanoff, *Macroscopic friction of dry granular materials*, Tribology series **43** (2004), 53–61.

[19] F. Da Cruz, S. Emam, M. Prochnow, J.N. Roux, F. Chevoir, *Rheophysics of dense granular materials: Discrete simulation of plane shear flows*, Phys. Rev. E **72** (2005), 021309.

[20] A. Daerr, S. Douady, *Two types of avalanches behaviour in granular media*, Nature **399** (1999), 241–43.

[21] A. Daerr, S. Douady, *Sensitivity of granular surface flows to preparation*, Europhys. Lett. **47** (1999b), 324–30.

[22] S. Deboeuf, E. Lajeunesse, O. Dauchot, B. Andreotti, *Flow rule, self-channelization and levees in unconfined granular flows*, Phys. Rev. Lett. **97** (2006), 158303.

[23] M. Depken, W. van Saaloos, M. van Hecke, *Continuum approach to wide shear zones in quasistatic granular matter*, Phys. Rev. E **73** (2006), 031302.

[24] D. Doppler, P. Gondret, T. Loiseleux, S. Meyer, M. Rabaud, *Relaxation dynamics of water-immersed granular avalanches*, J. Fluid Mech. **577** (2007), 161–181.

[25] S. Douady, B. Andreotti, A. Daerr, *On granular surface flow equations*, Eur. Phys. J. B **11** (1999), 131–42.

[26] S. Douady, B. Andreotti, P. Cladé, A. Daerr, *The four avalanche fronts; a test case for granular surface flow modeling*, Adv. Compl. Syst. **4** (2001), 509–22.

[27] D. Ertas, T.C. Halsey, *Granular gravitational collapse and chute flow*, Eurphys. Lett. **60** (2002), 931–937

[28] G. Félix, N. Thomas, *Relation between dry granular flow regimes and morphology of deposits: formation of levées in pyroclastic deposits*, Earth Planet. Sci. Lett. **221** (2004), 197–213.

[29] G. Felix, N. Thomas, *Evidence of two effects in the size segregation process in dry granular media*, Phys. Rev. E **70** (2005), 051307.

[30] D. Fenistein, J.W. van de Meent, M. van Hecke, *Universal and wide shear zones in granular bulk flow*, Phys. Rev. Lett. **92** (2004), 094301.

[31] Y. Forterre, O. Pouliquen, *Stability analysis of rapid granular chute flows: formation of longitudinal vortices*, J. Fluid Mech. **467** (2002), 361–87.

[32] Y. Forterre, O. Pouliquen, *Long-surface waves instability in dense granular flows*, J. Fluid Mech. **486** (2003), 21–50.

[33] Y. Forterre, *Kapiza waves give support for three-dimensional granular flow rheology*, J. Fluid Mech. **563** (2006), 123–132.

[34] Y. Forterre, O. Pouliquen, *Flows of dense granular media*, Annu. Rev. Fluid Mech. **40** (2008), 1–24.

[35] IA. Frigaard, C. Nouar, *On the usage of viscosity regularisation methods for viscoplastic fluid flow computation*, J. Non-Newt. Fluid Mech. **127** (2005), 1–26.

[36] Gdr MiDi, *On dense granular flows*, Eur. Phys. J. E **14** (2004), 341–65.

[37] J.D. Goddard, *Dissipative materials as constitutive Models for granular media*, Acta Mechanica **83** (1986), 3–13.

[38] I. Goldhirsch, *Rapid granular flows*, Ann. Rev. Fluid Mech. **35** (2003), 267–63.

[39] J.M.N.T. Gray, J.M. Wieland, K. Hutter, *Gravity-driven free surface flow of granular avalanches over complex basal topography*, Proc. R. Soc. Lond. A **455** (1999), 573–600.

[40] J.M.N.T. Gray, Y.C. Tai, S. Noelle, *Shock waves, dead zones and particle-free regions in rapid granular free-surface flows*, J. Fluid Mech. **491** (2003), 161–81.

[41] R. Greve, T. Koch, K. Hutter, *Unconfined flow of granular avalanches along a partly curved surface: experiments and theoretical predictions*, Proc. R. Soc. Lond. A **342** (1994), 573–600.

[42] P. Heinrich, A. Piatanesi, H. Hébert, *Numerical modelling of tsunami generation and propagation from submarine slumps; the 1998 Papua New Guinea event*, Geophys. J. Int. **145** (2001), 97–111.

[43] K.M. Hokanardottir, A.J. Hogg, *Oblique shocks in rapid granular flows*, Phys. Fluids **17** (2005), 077101.

[44] N. Huang, G. Ovarlez, F. Bertrand, S. Rodts, P. Coussot, D. Bonn, *Flow of wet granular materials*, Phys. Rev. lett **94** (2005), 028301.

[45] H.M. Jaeger, S.R. Nagel, R.P. Behringer, *Granular solids, liquids, and gases*, Rev. Mod. Phys. **68** (1996), 1259–73.

[46] J.T. Jenkins, S.B. Savage, *A theory for the rapid flow of identical, smooth, nearly elastic, spherical particles*, J. Fluid Mech. **130** (1983), 187–02.

[47] J.T. Jenkins, M.W. Richman, *Kinetic theory for plane flows of a dense gas of identical, rough, inelastic, circular disks*, Phys. Fluids **28** (1985), 3485–94.

[48] J.T. Jenkins, *Dense shearing flows of inelastic disks*, Phys. Fluids **18** (2006), 103307.

[49] P.C. Johnson, R. Jackson, *Frictional-collisional constitutive relations for granular materials, with application to plane shearing*, J. Fluid Mech. **176** (1987), 67–93.

[50] P. Jop, Y. Forterre, O. Pouliquen, *Crucial role of sidewalls in granular surface flows: consequences for the rheology*, J. Fluid Mech. **541** (2005), 167–92.

[51] P. Jop, Y. Forterre, O. Pouliquen, *A constitutive law for dense granular flows*, Nature **441** (2006), 727–30.

[52] P. Jop, *Hyddrodynamics modeling of granular flows in a modified Couette cell* Phys. rev. E **77** (2008), 032301.

[53] K. Kamrin, M.Z. Bazant, *A stochastic flow rule for granular materials*, cond-mat (2007), 0609448.

[54] D.V. Khakhar, A.V. Orpe, P. Andersen, J.M. Ottino, *Surface flow of granular materials: model and experiments in heap formation*, J. Fluid Mech. **441** (2001), 255–64.

[55] G. Hill, S. Yeung, S.A. Koehler, *Scaling vertical drag forces in granular media*, Europhys. Lett. **72** (1) (2005), 137–43.

[56] T.S. Komatsu, S. Inagaki, N. Nakagawa, S. Nasumo, *Creep motion in a granular pile exhibiting steady surface flow*, Phys. Rev. Lett. **86** (201), 1757–60.

[57] V. Kumaran, *The constitutive relation for the granular flow of rough particles, and its application to the flow down inclined plane*, J. Fluid Mech. **561** (2006), 1–42.

[58] L. Lacze, R.R. Kerswell, *Axisymmetric granular collapse: a transient 3D flow test of viscoplasticity* Phys. Rev. Lett. **102** (2009), 108305.

[59] E. Lajeunesse, A. Mangeney-Castelnau, J.P. Vilotte, 2004. *Spreading of a granular mass on a horizontal plane*, Phys. Fluids **16** (2004), 2371–81.

[60] E. Lajeunesse, J.B. Monnier, G.M. Homsy, *Granular slumping on a horizontal surface*, Phys. Fluids **17** (2005), 103302.

[61] A. Lemaitre, J.-N. Roux, F. Chevoir, *What do dry granular flows tell us about dense suspension rheology?*, Rheologica Acta (2009).

[62] P.A. Lemieux, D.J. Durian, *From avalanches to fluid flow: a continuum picture of grain dynamics down a heap*, Phys. Rev. Lett. **85** (2000), 4273–76.

[63] A.J. Liu, S.R. Nagel, *Jamming is not just cool any more*, Nature **396** (1998), 21–22.

[64] G. Lois, A. Lemaître, J. Carlson, *Numerical tests of constitutive laws for dense granular flows*, Phys. Rev. E **72** (2005), 051303.

[65] G. Lois, A. Lemaître, J.M. Carlson, *Emergence of multi-contact interactions in contact dynamics simulations of granular shear flows*, Europhys. Lett. **76** (2006), 318–24.

[66] M.Y. Louge, *Model for dense granular flows down bumpy inclines*, Phys. Rev. E **67** (2003), 061303.

[67] M. Louge, A. Valance, N. Taberlet, P. Richard, R. Delannay, *Volume fraction profile in channeled granular flows down an erodible incline*, Proc. Powders and Grains 2005 Stuttgart Rotterdam: A.A. Balkema.

[68] G. Lube, H.E. Huppert, R.S.J. Sparks, M. Hallworth, *Axisymmetric collapse of granular columns*, J. Fluid Mech. **508** (2004), 175–99.

[69] C.K.K. Lun, S.B. Savage, D.J. Jeffrey, N. Chepurniy, *Kinetic theories for granular flow: inelastic particles in Couette flow and slightly inelastic particles in a general flow field*, J. Fluid Mech. **140** (1984), 223–56.

[70] A. Mangeney-Castelnau et al., *On the use of Saint-Venant equations for simulating the spreading of a granular mass*, J. Geophys. Res. **110** (2005), B09103.

[71] A. Mangeney-Castelnau, F. Bouchut, N. Thomas, J.P. Vilotte, M.-O. Bristeau, *Numerical modeling of self-channeling granular flows and of their levee/channel deposits*, to appear in J. Geophys. Res.

[72] P. Mills, D. Loggia, M. Tixier, *Model for a stationary dense granular flow along an inclined wall*, Europhys. Lett. **45** (1999), 733–738.

[73] L.S. Mohan, K.K. Rao, P.R. Nott, *A frictional Cosserat model for the slow shearing of granular materials*, J. Fluid Mech. **457** (2002), 377–409.

[74] J.F. Morris, F. Boulay, *Curvilinear flows of noncolloidal suspensions: the role of normal stresses* J. Rheol. **43** (1999), 1213–1237.

[75] M. Naaïm, S. Vial, R. Couture, Saint-Venant approach for rock avalanches modelling. In *Multiple Scale Analyses and Coupled Physical Systems: Saint-Venant Symposium.* Presse de l'École Nationale des Ponts et Chaussées, Paris (1997).

[76] J.M. Ottino, D.V. Khakhar, *Mixing and segregation of granular materials*, Annu. Rev. Fluid Mech. **32** (2000), 55–91.

[77] M. Ouriemi, P. Aussillous, E. Guazzelli, *Bed-load transport by shearing flows*, J. Fluid Mech. **636** (2009), 295–319.

[78] M. Pailha, O. Pouliquen, *A two-phase flow description of the initiation of underwater granular avalanches*, J. Fluid Mech. **633** (2009), 115–135.

[79] O. Pouliquen, N. Renaut, *Onset of granular flows on an inclined rough surface: dilatancy effects*, J. Phys. II France **6** (1996), 923–35.

[80] O. Pouliquen, *Scaling laws in granular flows down rough inclined planes*, Phys. Fluids **11** (1999), 542–48.

[81] O. Pouliquen, *On the shape of granular fronts down rough inclined planes*, Phys. Fluids **11** (1999), 1956–1958.

[82] O. Pouliquen, Y. Forterre, *Friction law for dense granular flows: application to the motion of a mass down a rough inclined plane*, J. Fluid Mech. **453** (2002), 133–151.

[83] O. Pouliquen, F. Chevoir, *Dense flows of dry granular material*, C. R. Physique **3** (2002), 163–75.

[84] O. Pouliquen, *Velocity correlations in dense granular flows*, Phys. Rev. Lett. **93** (2004), 248001.

[85] O. Pouliquen, Y. Forterre, *A non local rheology for dense granular flows*, Phil. Trans. R. Soc. A in press (2009).

[86] L. Quartier, B. Andreotti, S. Douady, A. Daerr, *Dynamics of a grains on a sandpile model*, Phys. Rev. E **62** (2000), 8299–07.

[87] F. Radjai, S. Roux, *Turbulent-like behavior in quasi-static flow of granular media*, Phys. Rev. Lett. **89** (2003), 064302.

[88] J. Rajchenbach, *Dense, rapid flows of inelastic grains under gravity*, Phys. Rev. Let. **90** (2003), 144302.

[89] P. Rognon, J.-N. Roux, D. Wolf, M. Naaïm, F. Chevoir, *Rheophysics of cohesive granular materials*, Europhys. Lett. **74** (2005), 644–50.

[90] P. Rognon, J-N. Roux, M. Naaïm, F. Chevoir, *Dense flows of bidisperse assemblies of disks down inclined plane*, Phys. Fluids **19** (2007), 058101.

[91] J.-N. Roux, G. Combes, *Quasistatic rheology and the origin of strain*, C. R. Physique **3** (2002), 131–40.

[92] S. Roux, F. Radjai, Texture-dependent rigid-plasic behavior. In *Physics of dry granular media*, eds H.-J. Hermann et al., pp. 229–36. Kluwer Academic (1998).

[93] C. Ruyer-Quil, P. Manneville, *Improved modeling of flows down inclined planes*, Eur. Phys. J. B **15** (2000), 357–69.

[94] S.B. Savage, Granular flows down rough inclines: review and extension. In Mechanics of granular materials: new models and constitutive equations, Jenkins JT, Satake M. eds. (Elsevier, Amsterdam, 1983)

[95] S.B. Savage, M. Sayed, *Stresses developed by dry cohesionless granular materials sheared in an annular shear cell*, J. Fluid Mech. **142** (1984), 391–430.

[96] S.B. Savage, *The mechanics of rapid granular flows*, Adv. Appl. Mech. **24** (1984), 289–66.

[97] S.B. Savage, C.K.K. Lun, *Particle size. segregation in inclined chute flow of dry cohesionless granular solids*, J. Fluid Mech. **189** (1988), 311–35.

[98] S.B. Savage, K. Hutter, *The motion of a finite mass of granular material down a rough incline*, J. Fluid Mech. **199** (1989), 177–215.

[99] S.B. Savage, *Analyses of slow high-concentration flows of granular materials*, J. Fluid Mech. **377** (1998), 1–26.

[100] A. Schofield, P. Wroth, *Critical state soil mechanics*, London: McGraw-Hill (1968).

[101] L.E. Silbert, D. Ertas, G.S. Grest, T.C. Halsey, D. Levine, S.J. Plimpton, *Granular flow down inclined plane: Bagnold scaling and rheology*, Phys. Rev. E **64** (2001), 051302.

[102] L.E. Silbert, J.W. Landry, G.S. Grest, *Granular flow down a rough inclined plane: transition between thin and thick piles*, Phys. Fluids **15** (2003), 1–10.

[103] J.J. Stickel, R.L. Powell, *Fluid mechanics and rheology of dense suspensions*, Ann. Rev. Fluid Mech. **37** (2005), 129–49.

[104] N. Taberlet, P. Richard, A. Valance, R. Delannay, W. Losert, J.M. Pasini, J.T. Jenkins, *Super stable granular heap in thin channel*, Phys. Rev. Lett. **91** (2003), 264301.

[105] N. Taberlet, P. Richard, E. Henry, R. Delannay, *The growth of a super stable heap: an experimental and numerical study*, Europhys. Lett. **68** (4) (2004), 515–21.

[106] C.T. Veje, D.W. Howell, R.P. Behringer, *Kinematics of a two-dimensional granular couette experiment at the transition to shearing*, Phys. Rev. E **59** (1999), 739–45.

[107] D. Volfson, L.S. Tsimring, I.S. Aranson, *Order Parameter description of stationnary partially fluidized shear granular flows*, Phys. Rev. Lett. **90** (2003), 254301.

[108] J.M. Wieland, J.M.N.T. Gray, K. Hutter, *Channelized free-surface flow of cohesionless granular avalanches in a chute with shallow lateral curvature*, J. Fluid Mech. **392** (1999), 73–100.

[109] D.M. Wood, *Soil Behaviour and Critical State Soil Mechanics* Cambridge UK: Cambridge University Press (1990).

Yoël Forterre
Olivier Pouliquen
CNRS-Université de Provence
5, rue Enrico Fermi
F-13453 Marseille Cedex 13, France
e-mail: yoel.forterre@polytech.univ-mrs.fr
        Olivier.Pouliquen@univ-provence.fr

Glasses and Grains, 111–135
© 2011 Springer Basel AG

Theoretical Considerations for Granular Flow

Thomas C. Halsey

Abstract. Friction plays a key role in controlling the rheology of granular flows in both the "critical-state" and the "dense granular flow" regimes. Ertaş and Halsey, among others, have proposed that friction and inelasticity-enabled structures with a characteristic length scale in such flows can be directly linked to such rheologies, particularly that summarized in the "Pouliquen flow rule." In dense flows, "gear" states in which all contacts roll without frictional sliding are naively possible below critical coordination numbers. We construct an explicit example of such a state in $D = 2$; and show that organized shear can exist in this state only on scales $l < d/I$, where d is the grain size and I is the Inertial Number, characterizing the balance between inertial and pressure forces in the flow. Above this scale the packing is destabilized by centrifugal forces. Similar conclusions can be drawn in disordered packings of grains. We comment on the possible relationship between this length scale l and that which has been hypothesized to control the rheology.

1. Introduction

Flows of hard granular systems are ubiquitous in nature and technology, yet are still poorly understood. Compared to truly microscopic dynamical or statistical mechanical systems, an unusual feature of granular systems is that they are intrinsically frictional and dissipative. Not only can dissipation arise through inelastic collisions between the particles, but also from frictional sliding between smooth or rough grain surfaces.

In the quasi-static "critical state" regime [1], the strain rates are smaller than any time scale of the system. One would expect deformations in this regime to be predominantly plastic in nature, corresponding to relatively long-lived particle contacts on the time scale $\dot{\gamma}^{-1}$. Beyond this regime, there appears to be a "dense granular flow" regime, marked by strain rates obeying

$$\sqrt{P/\rho_g L^2} < \dot{\gamma} < \sqrt{P/\rho_g d^2}, \tag{1}$$

with P the pressure, ρ_g the density, L a characteristic flow scale, and d the grain diameter. It is suspected that this regime has a rheology that distinguishes it from a higher strain-rate regime, in which

$$\dot{\gamma} > \sqrt{P/\rho_g d^2}. \tag{2}$$

This higher-strain rate regime seems to be appropriately described by kinetic-theory based studies [2].

The degree to which such studies can be extended into the dense granular flow regime defined by Eq. (1) remains controversial. A number of authors have claimed that key aspects of the rheology of dense granular flow can be recovered within kinetic theory treatments that include particle inelasticity and interparticle friction [3]. Halsey and Ertaş, and Jenkins, have posited the appearance of coherent structures in dense granular flows, which are difficult to reconcile with the underlying assumptions of kinetic theories [4, 5]. A significant work by the group author GDR MiDi pointed out that much of the phenomenology of dense granular flows can be organized using the "Inertia Number" [6]

$$I \equiv d\dot{\gamma}/\sqrt{P/\rho_g}. \tag{3}$$

An interesting feature of these results is the anomalously large fluctuations in the grain velocities, which scale with I in the dense flow regime,

$$\frac{\langle (\vec{v}(\vec{x}) - \langle \vec{v}(\vec{x}) \rangle)^2 \rangle}{(\dot{\gamma}d)^2} \sim I^{-\chi}, \tag{4}$$

where in dimensionalities $D = 2, 3$, $\chi \approx 1$ [6, 7]. Setting

$$I \equiv d/\ell, \tag{5}$$

defines the scale ℓ of the coherent structures hypothesized by Halsey and Ertaş.

The motivation to consider the dimensionless parameter I as a ratio of the particle diameter and a "mesoscopic" length scale ℓ arises from the overall phase diagram of flow on an incline. The dominant observational fact about the steady-state incline flows is the "Pouliquen flow rule," which connects the average velocity u of a flow of height h with the height $h_{\text{stop}}(\theta)$ at which flow ceases for a chute of inclination θ [8]. [The angle of repose $\theta_{\mathcal{R}}(h)$ is the inverse of the function $h_{\text{stop}}(\theta)$.] The Pouliquen flow rule gives a scaling form for u,

$$\frac{u}{\sqrt{gh}} = F(\frac{h}{h_{\text{stop}}}), \tag{6}$$

with g the gravitational acceleration, and where the function $F(x) \approx 0.136x$ for glass beads, and has a similar linear form for other types of particles [8, 9, 10].

The scaling $u \propto h^{3/2}$ in the Pouliquen flow rule is consistent with the classical Bagnold rheology [11]. But the Pouliquen flow rule also connects the coefficient of this proportionality, $u = A_{\text{Bag}}(\theta)h^{3/2}$ with *the thickness of the pile at that inclination below which flow arrests.* Halsey and Ertaş have pointed out that this feature of the Pouliquen flow rule follows from the dependence of the rheology on I if it is assumed that flow is arrested for thicknesses less than $h_{\text{stop}} \sim d/I = \ell$ [4]. Although Halsey and Ertaş proposed that this length scale corresponded to large-scale "eddy" structures in the flow, such eddies have proven elusive; a direct attempt to measure ℓ by measuring velocity correlations in flows was not successful [7].

The broad features of Pouliquen's conclusions have been confirmed by a series of numerical studies [9]. For relatively thin piles, the Bagnold rheology breaks down, but the thicker piles show a Bagnold rheology and obey the Pouliquen flow rule, albeit with a slightly larger value of β (the crossover is examined numerically in [12]). Although at large, but not enormous, values of particle stiffness, instantaneous coordination numbers are high (unlike in kinetic theory treatments), the duration of two-body collisions is still short compared to inverse strain rates $\dot{\gamma}^{-1}$ [13]. The density in the interior of the piles is independent of depth.

In this article I turn to the role of friction at enduring contacts in plastic flows. Of course, assuming that such long-lived contacts persist into the dense granular flow regime is problematic, due to the considerations regarding the short duration of two-body collisions mentioned above. To extend the conclusions of this work into this regime will require that the correlations characteristic of plastic flow with rolling and sliding contacts survive in a regime of frequent, albeit short, collisions between any particle and its neighbors. This would imply that a full theoretical description of this regime will require a synthesis of plastic flow and kinetic theory concepts.

Provided that the microscopic coefficient of friction μ defining the maximum value of the tangential force T at a contact divided by the normal force N is appreciable, we would anticipate that a significant fraction of long-lived contacts in a deforming granular packing will be rolling as opposed to sliding. Simple counting of the number of constraints vs the number of variables indicates that states in which all contacts are rolling are not possible at coordination numbers $Z_c > 3$ for $D = 2$ or $Z_c > 4$ for $D = 4$. These results are simple extensions to the dynamical case of the famous "isostatic" criteria for static packings [14].

An interesting soluble case in $D = 2$ is presented by the honeycomb lattice, for which all particles have a coordination number $Z_c = 3$ exactly. This allows for a general solution of all states obeying the rolling constraint, corresponding to a full solution of the kinematics for this packing, in the limit $\mu \to \infty$, over times short enough that collisions do not degrade the lattice. It is also possible to solve exactly for both the tangential and normal forces exerted between the particles in this lattice; I present this explicitly for a particularly symmetric case of motion.

A striking feature of the kinematic solution is that even if the overall shear remains moderate, the rotational velocities of the particles are quite large. For a region of the packing of size ℓ with a constant shear $\dot{\gamma}$, the typical rotational velocity of the particles is $\omega \sim \dot{\gamma}\ell/d$. In addition, forces develop that limit the size of these regions of constant shear to

$$\ell < \frac{\sqrt{P/\rho_g}}{\dot{\gamma}}, \tag{7}$$

due to the requirement that all normal forces remain compressive. (Dry granular packings are unable to support tensions between the particles.) Equation (7) corresponds to the definition of the coherent structure size ℓ given in Eqs. (3), (5) above.

Turning to disordered lattices, we see that the rotations arise from an underlying short-ranged anti-ferromagnetic ordering in the rotational velocities of the individual grains. This short-ranged ordering is frustrated by the existence in disordered lattices of "odd plaquettes" with odd numbers of links around the plaquette. In fact, it follows that it is not possible, even for quite large values of the coefficient of friction, for all of the particle contacts in a lattice with odd plaquettes to be rolling contacts. However, if sliding contacts are allowed, such that every odd plaquette has at least one sliding contact, then the remaining contacts can be rolling contacts; results on spin glasses then imply that it is possible that long-range ordered antiferromagnetic "spin-glass" regions will form. Of course, given the lack of direct correspondence between a statistical mechanical model (spin glasses) and a driven dissipative non-equilibrium system, it is not possible to draw rigorous conclusions from this analogy.

The remainder of this article is divided into four sections. In Section 2, we derive the critical coordination numbers for rolling states, and we solve the rolling kinematics of the honeycomb lattice. In Section 3, we solve a special case for the dynamics of this state. In Section 4, we turn to disordered lattices in $D = 2$, using a Fokker-Planck approach to show the basis for the anti-ferromagnetic ordering and the analogy to spin glasses. We also comment on the applicability of these results to the three-dimensional case. In Section 5, we conclude.

2. Kinematics of rolling states

Consider a set of N_g spherical grains of diameter d, indexed by i, in D dimensions with velocities $\vec{v}_i$ and rotational velocities (in $D = 2$ or $D = 3$, the generalization to higher dimensions is obvious) $\vec{\omega}_i$. We further suppose the existence of pairs of grains $\langle ij \rangle$ in contact with one another, and for which there is no relative motion of the surface points in contact (corresponding to frictional locking of the particles). Suppose that the vector connecting the pair $\langle ij \rangle$ is $\delta \vec{w}_{\langle ij \rangle}$, with $|\delta \vec{w}_{\langle ij \rangle}| = d$. Then the requirement that the relative motion of the surface points in contact be zero is

$$\vec{v}_i - \vec{v}_j \equiv \delta \vec{v}_{\langle ij \rangle} = \frac{1}{2}(\vec{\omega}_i + \vec{\omega}_j) \times \delta \vec{w}_{\langle ij \rangle}. \tag{8}$$

Taking the derivative with respect to time of this constraint yields a constraint for the accelerations of the grains $\vec{a}_i$ and angular accelerations $\vec{\Gamma}_i$:

$$\delta \vec{a}_{\langle ij \rangle} = \frac{1}{2}(\vec{\omega}_i + \vec{\omega}_j) \times \delta \vec{v}_{\langle ij \rangle} + \frac{1}{2}(\vec{\Gamma}_i + \vec{\Gamma}_j) \times \delta \vec{w}_{\langle ij \rangle}. \tag{9}$$

These equations substantially constrain both the motion of the grains and the forces between the grains. Equation (8) gives $D-1$ constraints per contact. The requirement that the grains stay in contact gives one further constraint. Since each particle has $D(D+1)/2$ degrees of freedom without contacts, this means that the total number of degrees of freedom N_F is

$$\frac{N_F}{N_g} = \frac{D}{2}(D+1-Z_c), \tag{10}$$

where Z_c is the average coordination number of the grains. We thus see that the average coordination $Z_c \leq D+1$ in order for the rolling state to be mobile at all, and Eq. (10) then gives the effective number of degrees of freedom remaining to the packing.

Note that the coordination number can exceed this value if some of the contacts are sliding instead of rolling. In this case, there are D constraints per rolling contact, and only one contraint at a sliding contact. If we write the average number of rolling and sliding contacts per particle respectively as Z_C^R and Z_C^S, then[16]

$$\frac{N_F}{N_g} = \left[\frac{D}{2}(D+1) - \frac{D}{2}Z_c^R - \frac{1}{2}Z_C^S \right], \tag{11}$$

which is similar to an "isostatic" argument for a packing with a mixture of Coulomb saturated and unsaturated contacts [17].

The forces are even more completely determined. The total number of contact forces is exactly the same as the number of constraints on the accelerations given by Eq. (9), with the result that all of these forces are determined by the contact network, the velocities and angular velocities of the particles, and the boundary conditions, even for $Z_c < D+1$. (This result can be extended to the case of sliding contacts [18].)

With the kinematics thus determined, the equations of motion of the particles can be integrated until one of two possible types of crisis occurs to disrupt the network (See Figure 1).

1. If two grains collide, they will very rapidly (within a collision time set by the Young's modulus of the particles) establish a contact with a finite normal force, provided they are sufficiently inelastic. This also changes the network, and hence the kinematics. In this case the impulse arising from the collision will, in general, result in a jump in the velocities and forces, which may be significant in the neighborhood of the new contact. Corresponding to the change in kinetic energy caused by the velocity jumps there will be a net energy dissipation corresponding to the collision event [19].

2. If the normal force between two grains becomes zero, the grains will separate. This effectively changes the contact network, and the kinematics must now

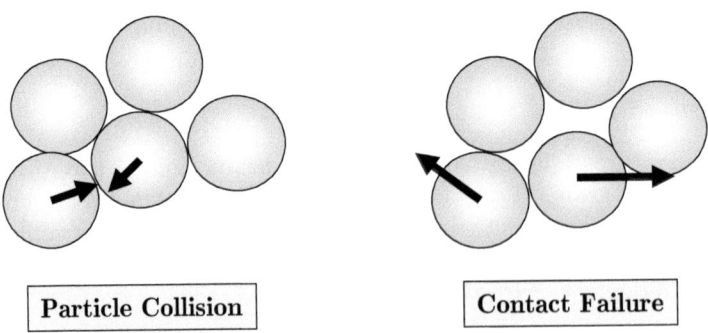

FIGURE 1. Two types of crisis can disrupt a lattice of rolling grains: a) two grains collide, creating a new contact, or b) two grains separate due to the normal force between them reaching zero.

be solved with the new contact network. However, we do not anticipate that this change in kinematics will correspond to any jump in the velocities or forces. This event does not lead to any dissipation.

In a statistical steady state, the number of contact failures per unit time should equal the number of new contacts created by collisions per unit time.

2.1. The honeycomb lattice

In $D = 2$, we can construct an explicit example of a rolling state at the critical coordination number $Z_c = 3$ by considering a honeycomb lattice (Figure 2). We consider a relatively symmetric lattice, characterized by one angle θ; the basic plaquettes of the lattice are equilateral (although not necessarily regular) hexagons.

We describe the system as a lattice of "doublets" made up of two grains in contact. The positions of these doublets we index by (j, k), with either j, k both even, or j, k both odd. The positions of the centers of the doublets are then

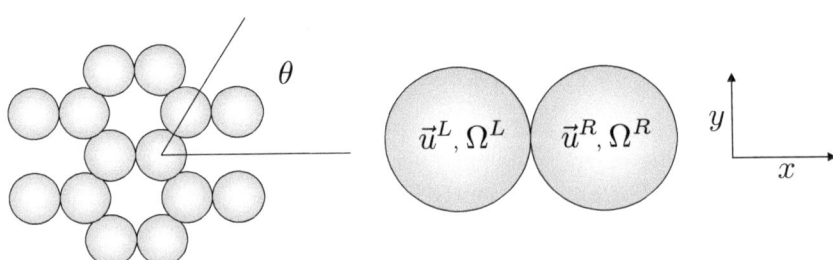

FIGURE 2. A honeycomb lattice has the critical coordination number $Z_c = 3$ at which a frictional packing becomes mobile in $D = 2$. We work with a set of honeycomb lattices characterized by the single parameter θ. We also show the fundamental "doublet" and the coordinate system.

$(x, y) = (c_x j, c_y k)$, with $c_x = d(1 + \cos\theta)$ and $c_y = d\sin\theta$. Note that the axes of the Bravais lattice of doublets are oriented along the angles $\pm\theta/2$. With the doublets oriented along the x-axis, the requirement that the two grains stay in contact and have a rolling contact constrains the velocities and rotational velocities of the left-hand and right-hand grains, $u_x^L, u_y^L, \Omega^L; u_x^R, u_y^R, \Omega^R$,

$$u_x^L = u_x^R \equiv u_x, \quad u_y^L \equiv u_y - \tfrac{d}{4}(\Omega^L + \Omega^R), \quad u_y^R \equiv u_y + \tfrac{d}{4}(\Omega^L + \Omega^R) , \quad (12)$$

which reduces the six degrees of freedom of the two grains taken independently to the four degrees of freedom remaining after the imposition of the constraint that the contact be a permanent, rolling contact. These four degrees of freedom are conveniently taken as the average x and y velocities of the doublet, u_x, u_y, and two (re-dimensioned) combinations of the two rotational velocities $\Delta = \tfrac{d}{2}(\Omega^L + \Omega^R)$, $\xi = \tfrac{d}{2}(\Omega^L - \Omega^R)$.

Now let us consider the constraints imposed by the requirement that the contacts between the doublet and other grains be rolling. Consider the contacts between the doublet at j, k and the doublets at $j-1, k-1$ and $j+1, k-1$. Let us write the vector $z(j, k)$ desciribing the doublet kinematical state,

$$z(j, k) = (u_x(j, k), u_y(j, k), \Delta(j, k), \xi(j, k)). \quad (13)$$

The contact rolling equations have the form

$$z(j, k) = \bar{A}_- z(j-1, k-1) + \bar{A}_+ z(j+1, k-1), \quad (14)$$

with $\bar{A}_-(\theta)$ and $\bar{A}_+(\theta)$ given, after some labor, by

$$\bar{A}_- = \begin{pmatrix} \tfrac{1}{2} & \tfrac{1}{2}\tan(\tfrac{\theta}{2}) & 0 & \tfrac{1}{4}\tan(\tfrac{\theta}{2}) \\ \tfrac{\cot(\theta)}{2} & \tfrac{1}{2} & \tfrac{1}{4} & 0 \\ 0 & -\tfrac{1}{1+\cos\theta} & -\tfrac{1}{2} & \tfrac{1}{2}\tfrac{\cos\theta}{1+\cos\theta} \\ \csc\theta & 0 & -\tfrac{1}{2} & \tfrac{1}{2} \end{pmatrix}, \quad (15)$$

and

$$\bar{A}_+ = \begin{pmatrix} \tfrac{1}{2} & -\tfrac{1}{2}\tan(\tfrac{\theta}{2}) & 0 & -\tfrac{1}{4}\tan(\tfrac{\theta}{2}) \\ -\tfrac{\cot(\theta)}{2} & \tfrac{1}{2} & -\tfrac{1}{4} & 0 \\ 0 & \tfrac{1}{1+\cos\theta} & -\tfrac{1}{2} & -\tfrac{1}{2}\tfrac{\cos\theta}{1+\cos\theta} \\ -\csc\theta & 0 & \tfrac{1}{2} & \tfrac{1}{2} \end{pmatrix}. \quad (16)$$

Solutions to Eq. (14) can be written in the Bloch-Floquet form

$$z(j, k) = \int_{-\pi/2}^{\pi/2} \frac{dq}{\pi} e^{iqj} \sum_{n=1}^{4} \alpha_n(q)(\lambda_n(q))^k v_n(q), \quad (17)$$

with

$$\bar{A}(q)v_n(q) = \lambda_n(q)v_n(q), \quad (18)$$

where $\bar{A}(q)$ is

$$\bar{A}(q) = \cos(q) \begin{pmatrix} 1 & 0 & 0 & 0 \\ 0 & 1 & 0 & 0 \\ 0 & 0 & -1 & 0 \\ 0 & 0 & 0 & 1 \end{pmatrix}$$

$$+ \, \imath \sin(q) \begin{pmatrix} 0 & -\tan(\frac{\theta}{2}) & 0 & -\frac{1}{2}\tan(\frac{\theta}{2}) \\ -\cot(\theta) & 0 & -\frac{1}{2} & 0 \\ 0 & \frac{2}{1+\cos\theta} & -1 & -\frac{\cos\theta}{1+\cos\theta} \\ -2\csc\theta & 0 & 1 & 0 \end{pmatrix}.$$

We can now directly diagonalize $\bar{A}(q)$ to obtain the eigenvalues and eigenvectors.

$$\{\lambda_1(q); v_1\} = \{-1; [0, -1/2, \imath\cot(q/2), 1]\}, \tag{19}$$

$$\{\lambda_2(q); v_2\} = \{1; [0, -1/2, -\imath\tan(q/2), 1]\}, \tag{20}$$

$$\{\lambda_3(q); v_3\} = \{e^{-\imath q}; [\frac{\sin\theta}{2}, \frac{\cos\theta}{2}, 0, 1]\}, \tag{21}$$

$$\{\lambda_4(q); v_4\} = \{e^{\imath q}; [-\frac{\sin\theta}{2}, \frac{\cos\theta}{2}, 0, 1]\}. \tag{22}$$

These modes can obviously be combined to give a wide variety of different possible motions. We are most interested, however, in motions corresponding to low wavenumber deformations of the lattice. With some manipulations, we can determine the general solution for constant velocity gradients. With $\beta_{1,2,3,4} \in \Re$, we write

$$z(j,k) = \begin{pmatrix} \frac{\sin\theta}{2}(\beta_3(j-k) - \beta_4(j+k)) \\ \frac{\cos\theta}{2}(\beta_3(j-k) + \beta_4(j+k)) - \frac{\beta_2}{2}j + \frac{\beta_1}{4}(-1)^k \\ \beta_1 j(-1)^k \\ \beta_2 j + \beta_3(j-k) + \beta_4(j+k) - \frac{\beta_1}{2}(-1)^k \end{pmatrix}. \tag{23}$$

Note that for $\beta_1 \neq 0$, but $\beta_{2,3,4} = 0$, there is no large-scale motion of the lattice, but rather compensating motions of neighboring doublets. Although we have not introduced the concept of an excitation energy, the β_1 mode is reminiscent of "optical" modes in standard lattice dynamics, which have finite energy even at zero wave-vector, and corresponds similarly to compensating motions inside a Bravais lattice cell.

The other three modes correspond to simple shear motions. The β_2 mode corresponding to shear along the y-direction (Figure 3), while the β_3 and β_4 modes corresponding to flow perpendicular to the directions $\pm\theta$ (see Figure 4). Flows such as extensional flows that combine simple shear motions can be easily constructed from these three simple shearing modes.

It is notable that in all of the simple shearing modes, the parameter $\xi = \frac{d}{2}(\Omega^L - \Omega^R)$ increases linearly across the shearing region (Figure 3). While the

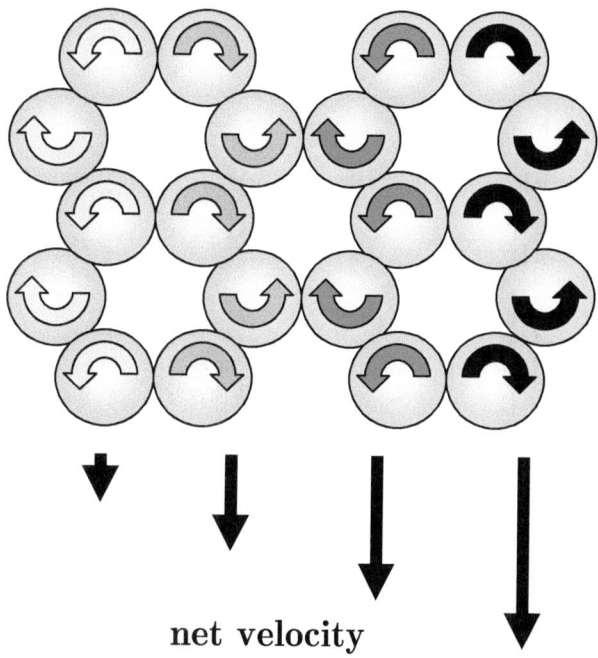

net velocity

FIGURE 3. The β_2 mode from Eq. (23) corresponds to grains counter-rotating in each vertical chain, with the magnitude of the rotations increasing linearly with the coordinate x. This generates an overall pure shear, with the velocity in the vertical direction.

average angular velocity $\langle \omega \rangle$ obeys

$$\langle \omega \rangle \sim \dot{\gamma}, \tag{24}$$

where $\dot{\gamma}$ is the overall shear, the average value of the angular velocity squared, $\langle \omega^2 \rangle$ depends additionally on the size of the system (or of the coherently shearing region within the system) ℓ,

$$\langle \omega^2 \rangle \sim \left(\frac{\dot{\gamma} \ell}{d} \right)^2. \tag{25}$$

3. Forces in rolling states

Simply showing that a state is kinematically possible does not imply that it is dynamically feasible – to accomplish this latter, we must also determine a set of interparticle forces with which it is consistent. For a granular system, this will be a set of normal forces N and tangential forces T across each contact, subject to the two constraints that $N > 0$, since grains cannot exert tensional forces on one another, and $T \leq \mu N$, with μ the coefficient of friction.

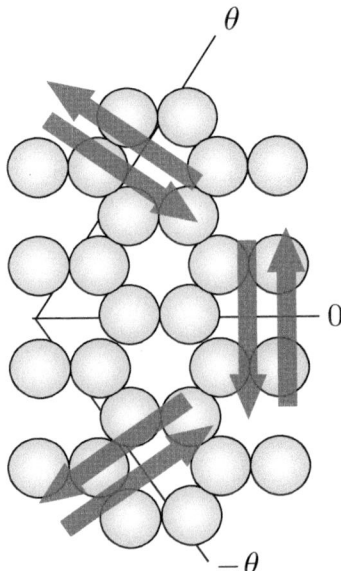

FIGURE 4. The three acoustic modes of the kinematics correspond to shear directed perpendicular to the y-axis, and perpendicular to the directions $\pm\theta/2$.

Note that despite the fact that we are considering rolling contacts, we assume that any value of $T \leq \mu N$ is admissable, and do not restrict ourselves to values of T corresponding to "free rolling" [19]. This is due to the fact that our particles are in general constrained by several contact forces, so the contact dynamics will be governed by the laws of "tractive rolling", which allow any value of $T \leq \mu N$.

Returning to our honeycomb lattice, consider a doublet at position (j, k). Such a doublet is influenced by eight forces exerted across its four contacts, as well as by two internal forces exerted by the two grains on one another. It is convenient to use a geographical notation to describe the external forces – the forces exerted by the doublet at $(j-1, k+1)$ on the doublet at (j, k) are termed N_{NW}, T_{NW}, with T defined so that a force in the counterclockwise direction is positive. Similarly, we define the forces exerted by the $(j+1, k+1)$, $(j+1, k-1)$ and $(j-1, k-1)$ doublets respectively as (N_{NE}, T_{NE}), (N_{SE}, T_{SE}), and (N_{SW}, T_{SW}) (see Figure 5). The force exerted at the internal contact is (N_0, T_0), with T_0 defined so that $T_0 > 0$ corresponds to a counterclockwise force on each grain.

With some tedious but straightforward algebra, we can convert the six equations of motion of the two particles in the doublet into four equations coupled to our natural kinematical variables, plus two equations determining the internal forces N_0, T_0 (which we here omit). We write M and I_M for the masses and moments of inertia of the grains, with $\kappa = I_M/Md^2$. The result for the dynamical

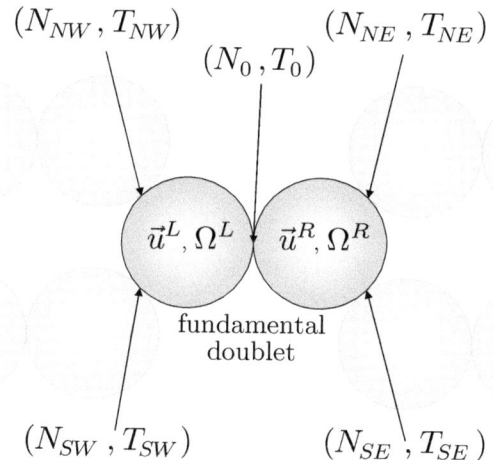

FIGURE 5. Eight forces are exerted on a fundamental doublet of the honeycomb lattice by its neighbors; two additional forces are exerted by the doublet particles on one another.

equation for the doublet is [15]

$$\begin{pmatrix} N_{NW} \\ T_{NW} \\ N_{NE} \\ T_{NE} \end{pmatrix} = \bar{B}(\theta) \begin{pmatrix} N_{SW} \\ T_{SW} \\ N_{SE} \\ T_{SE} \end{pmatrix} + 2M\bar{C}(\theta) \begin{pmatrix} \dot{u}_x \\ \dot{u}_y \\ (1+\kappa)\dot{\Delta} \\ \kappa\dot{\xi} \end{pmatrix}, \qquad (26)$$

with $\bar{B}(\theta)$ and $\bar{C}(\theta)$ given by

$$\bar{B}(\theta) = \begin{pmatrix} 2\sin^2(\theta/2) & -\sin\theta & \cos\theta & -\sin\theta \\ \sin\theta - \tan(\theta/2) & -\cos\theta & -\sin\theta + \tan(\theta/2) & 2\sin^2(\theta/2) \\ \cos\theta & \sin\theta & 2\sin^2(\theta/2) & \sin\theta \\ \sin\theta - \tan(\theta/2) & 2\sin^2(\theta/2) & -\sin\theta + \tan(\theta/2) & -\cos\theta \end{pmatrix}, \qquad (27)$$

and

$$\bar{C}(\theta) = \frac{1}{2} \begin{pmatrix} 1 & -\csc\theta & \tan(\theta/2) & -\cot\theta \\ -\tan(\theta/2) & 0 & \frac{\cos\theta}{1+\cos\theta} & 1 \\ -1 & -\csc\theta & -\tan(\theta/2) & -\cot\theta \\ -\tan(\theta/2) & 0 & \frac{\cos\theta}{1+\cos\theta} & -1 \end{pmatrix}, \qquad (28)$$

Note that $(N_{SW}(j,k), T_{SW}(j,k)) = (N_{NE}(j-1, k-1), T_{NE}(j-1, k-1))$, with equivalent identities for the other forces. The forces can be written as

$$\begin{pmatrix} N_{NW}(j,k) \\ T_{NW}(j,k) \\ N_{NE}(j,k) \\ T_{NE}(j,k) \end{pmatrix} = \int_{-\pi/2}^{\pi/2} \frac{dq}{\pi} e^{iqj} \begin{pmatrix} N_{NW}(q,k) \\ T_{NW}(q,k) \\ N_{NE}(q,k) \\ T_{NE}(q,k) \end{pmatrix}. \qquad (29)$$

We then can immediately write the equation determining solutions to the homogeneous problem,

$$
\begin{pmatrix} N_{NW}(q,k) \\ T_{NW}(q,k) \\ N_{NE}(q,k) \\ T_{NE}(q,k) \end{pmatrix} = \bar{B}(\theta)\bar{D}(q) \begin{pmatrix} N_{NW}(q,k-1) \\ T_{NW}(q,k-1) \\ N_{NE}(q,k-1) \\ T_{NE}(q,k-1) \end{pmatrix},
\tag{30}
$$

with

$$
\bar{D}(q) = \begin{pmatrix} 0 & 0 & e^{-\imath q} & 0 \\ 0 & 0 & 0 & e^{-\imath q} \\ e^{\imath q} & 0 & 0 & 0 \\ 0 & e^{\imath q} & 0 & 0 \end{pmatrix}.
\tag{31}
$$

The eigenvalues and eigenvectors $\{\nu_n; u_n\}$ of $\bar{\mathcal{B}}(\theta, q) = \bar{B}(\theta)\bar{D}(q)$ are given by

$$
\begin{aligned}
\{\nu_1(q); u_1\} &= \{-1; [e^{-\imath q}\tan\theta, e^{-\imath q}, -\tan\theta, 1]\}, & (32) \\
\{\nu_2(q); u_2\} &= \{1; [-e^{-\imath q}\tan\theta, -e^{-\imath q}, -\tan\theta, 1]\}, & (33) \\
\{\nu_3(q); u_3\} &= \{e^{-\imath q}; [0, 0, \cot(\theta/2), 1]\}, & (34) \\
\{\nu_4(q); u_4\} &= \{e^{\imath q}; [-\cot(\theta/2), 1, 0, 0]\}. & (35)
\end{aligned}
$$

These eigenvectors are in fact quite intuitive. Each of the eigenvectors 2–4 corresponds to a set of parallel force chains in one direction in the lattice, as is illustrated in Figure 6; eigenvector 1 is more complex.

Given the homogeneous solutions, it is straightforward to determine the solution to the inhomogeneous problem for which the packing is moving. Writing

$$
2M\bar{C}(\theta) \begin{pmatrix} \dot{u}_x(q,k) \\ \dot{u}_y(q,k) \\ (1+\kappa)\dot{\Delta}(q,k) \\ \kappa\dot{\xi}(q,k) \end{pmatrix} = \sum_{n=1}^{4} \upsilon_n(q,k)u_n(\theta,q),
\tag{36}
$$

with

$$
\upsilon_n(j,k) = \int_{-\pi/2}^{\pi/2} \frac{dq}{\pi} e^{\imath q j} \upsilon_n(q,k),
\tag{37}
$$

we have the inhomogeneous equation

$$
\begin{pmatrix} N_{NW}(q,k) \\ T_{NW}(q,k) \\ N_{NE}(q,k) \\ T_{NE}(q,k) \end{pmatrix} = \bar{B}(\theta,q) \begin{pmatrix} N_{NW}(q,k-1) \\ T_{NW}(q,k-1) \\ N_{NE}(q,k-1) \\ T_{NE}(q,k-1) \end{pmatrix} + \sum_{n=1}^{4} \upsilon_n(q,k)u_n.
\tag{38}
$$

This equation is easy to solve. Writing

$$
\begin{pmatrix} N_{NW}(q,0) \\ T_{NW}(q,0) \\ N_{NE}(q,0) \\ T_{NE}(q,0) \end{pmatrix} = \sum_{n=1}^{4} \tau_n(q)u_n,
\tag{39}
$$

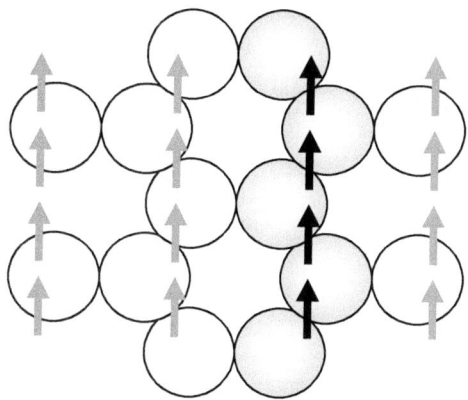

FIGURE 6. The fundamental homogeneous modes of the forces correspond principally to force chains, including both normal and tangential
forces, directed along the lattice directions of the honeycomb lattice. The
force chains shown correspond to the second force eigenvector, given in
Eq. (33).

we immediately obtain

$$
\begin{pmatrix} N_{NW}(q,k) \\ T_{NW}(q,k) \\ N_{NE}(q,k) \\ T_{NE}(q,k) \end{pmatrix} = \sum_{n=1}^{4} \left[\sum_{k'=0}^{k-1} v_n^{k'} v_n(q,k-k') + v_n^k \tau_n(q) \right] u_n. \tag{40}
$$

As an example, let us consider extensional flow: $u_x \propto x$, $u_y \propto -y$. This
type of flow has the advantage that it preserves the overall symmetries of the
honeycomb lattice, and can thus be interpreted parametrically by making the
angle θ a function of time, $\theta(t)$. Let us choose the simple case that

$$
\theta(t) = -\Lambda t + \theta_0 \tag{41}
$$

with $\Lambda > 0$.

With some labor, it is possible to determine the inhomogeneous contribution
to the forces corresponding to this motion,

$$
\begin{pmatrix} N_{NW}(j,k) \\ T_{NW}(j,k) \\ N_{NE}(j,k) \\ T_{NE}(j,k) \end{pmatrix}_{\text{inhom.}}
$$

$$
= \frac{dM\Lambda^2}{2} [\cos\theta \tan(\theta/2)] \left\{ \frac{k^2}{2} \begin{pmatrix} \tan\theta \\ 1 \\ \tan\theta \\ -1 \end{pmatrix} + jk \begin{pmatrix} -\cot(\theta/2) \\ 1 \\ \cot(\theta/2) \\ 1 \end{pmatrix} \right\} \tag{42}
$$

where we have neglected terms of less than quadratic order in j, k.

It is clear that this inhomogeneous contribution includes negative values of the normal forces at sufficiently low values of k/j, and that these scale with k^2 (at fixed k/j). In addition, it is also easy to determine the inhomogeneous part of the internal normal force within a doublet, $N_{0,\text{inhom.}}$ (at quadratic order in j, k) [15],

$$N_{0,\text{inhom.}} = -\frac{dM\Lambda^2 \sin\theta}{2} k^2 \tag{43}$$

which is also negative, and also scales with k^2.

Thus, no matter what the value of the homogeneous term for the forces (corresponding to the term proportional to τ in Eq. (40)), the requirement that normal forces be positive will ultimately be overwhelmed by the growth of the inhomogeneous term. Note that it is the contacts most perpendicular to the direction of acceleration that are most at risk of losing their normal forces. In addition, with a finite coefficient of friction, the tangential forces will ultimately saturate the Coulomb criterion and slippage will occur at grain contacts, invalidating our kinematical assumptions. However, even with an *infinite* coefficient of friction, the motion of the packing is limited to a domain size controlled by the requirement that all normal forces be compressional. If the scale of the homogeneous forces is determined by the average pressure in the packing P, then the scale ℓ of this domain size is determined by

$$\frac{\ell}{d} = \frac{1}{\Lambda d}\sqrt{P/\rho_g}, \tag{44}$$

with the density $\rho_g \sim M/d^2$. Comparing to Eq. (3), we see that this is equivalent to

$$\ell = \frac{d}{I}, \tag{45}$$

which connects the length scale over which frictionally dominated motion can determine the kinematics with the Inertia Number I, in a manner consistent with the physical picture of Halsey and Ertaş.

Note that this result can also be obtained on dimensional grounds. The accelerations in a shearing state, as determined in the above calculation, will typically be directed normal to the contacts, leading to a local relative acceleration of

$$\vec{a}_i - \vec{a}_j \sim \dot{\gamma}^2 d. \tag{46}$$

However, if the motions are coherent, as in a shearing motion, then accelerations will accumulate across a region ℓ, so that the typical acceleration in this region will be

$$\vec{a}_i \sim \dot{\gamma}^2 \ell. \tag{47}$$

Similarly, the forces will need to accumulate to drive these accelerations, so we conclude that the force scale will be

$$N \sim M\frac{(\dot{\gamma}\ell)^2}{d} \tag{48}$$

which is the scale of the inhomogenenous forces computed explicitly above. This follows entirely from the coherent nature of the accelerations, and does not require that the contacts be rolling. Thus, any state in which Eqs. (46)–(48) hold will also have its regions of coherent shear limited to the scale $\ell = d/I$, including much more disorderly states than the honeycomb lattice. I will return to this point in Section 4 below.

4. Kinematics of random grain packings

Let us reconsider the kinematics of the honeycomb lattice state, as defined by Eq. (17). An initially surprising feature of this state is the role played by $\xi = \frac{d}{2}(\Omega^L - \Omega^R)$, which increases linearly with distance in states with overall linear behavior of the average velocities u_x and u_y. Thus, one way to view the motion of the honeycomb state is by decomposing it into alternating sub-lattices, A and B (see Figure 7), on which the particles are approximately counter-rotating with respect to one another. The spatial variations of these counter-rotations then determine the overall spatial structure of the flow. Now we must consider random lattices, and determine, in particular, if there is any remnant of this feature for such lattices.

To determine the nature of the random lattice solution that respects the constraining large-scale motions, we apply a two-point Fokker-Planck approximation. Consider a random walk that moves entirely between grains in contact with one another (see Figure 8). Provided that the instantaneous contact network percolates, such a random walk can access arbitrarily distantly separated parts of the grain packing. Over the set of all particle contacts $\langle ij \rangle$ we can define a probability distribution $\rho(\theta, \delta \vec{v})$ on the angles of the contacts θ_{ij} and the differences in grain velocity across the contacts $\delta \vec{v}_{ij} = \vec{v}_i - \vec{v}_j$. For simplicity of notation we

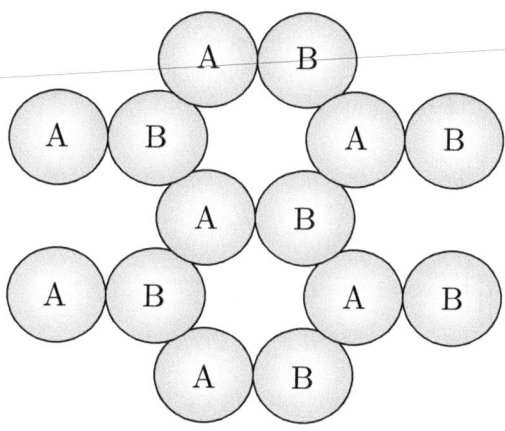

FIGURE 7. The honeycomb lattice shear solutions correspond to approximate counter-rotation of two alternating sublattices of grains, here indexed by A, B.

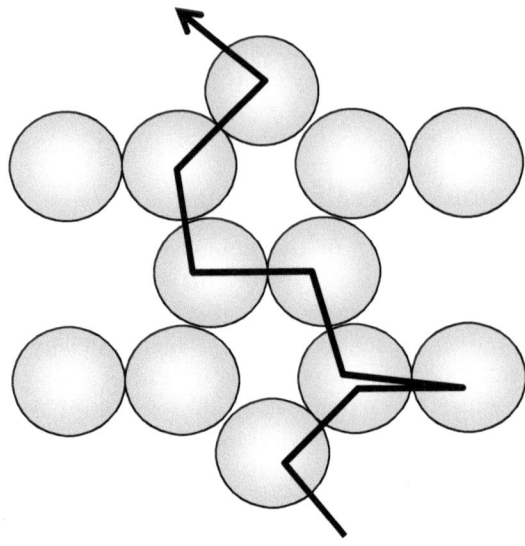

FIGURE 8. In the Fokker-Planck approximation, we consider random walks from grain to grain passing always through contacts between the grains.

are confining ourselves to two dimensions; however, the generalization to arbitrary dimensions is clear, and is assumed in much of the discussion below.

We now assume that in the random walk described in the previous paragraph, subsequent steps in the random walk correspond to uncorrelated choices of $\theta_{ij}, \delta \vec{v}_{ij}$ from the distribution $\rho(\theta, \delta \vec{v})$. (The neglect of correlations in this approximation is clearly dangerous, particularly in lower dimensionalities, of which more later.)

Let us define a matrix corresponding to the average changes in position and velocity corresponding to one step in this random walk,

$$\bar{\mathcal{M}} = \begin{pmatrix} \langle \delta w_x \delta w_x \rangle & \langle \delta w_x \delta w_y \rangle & \langle \delta w_x \delta v_x \rangle & \langle \delta w_x \delta v_y \rangle \\ \langle \delta w_y \delta w_x \rangle & \langle \delta w_y \delta w_y \rangle & \langle \delta w_y \delta v_x \rangle & \langle \delta w_y \delta v_y \rangle \\ \langle \delta v_x \delta w_x \rangle & \langle \delta v_x \delta w_y \rangle & \langle \delta v_x \delta v_x \rangle & \langle \delta v_x \delta v_y \rangle \\ \langle \delta v_y \delta w_x \rangle & \langle \delta v_y \delta w_y \rangle & \langle \delta v_y \delta v_x \rangle & \langle \delta v_y \delta v_y \rangle \end{pmatrix}. \tag{49}$$

In this matrix, the averages are easily written, e.g.,

$$\langle \delta w_x \delta w_x \rangle = \int d\theta \int d(\delta \vec{v}) \rho(\theta, \delta \vec{v}) (d \cos \theta)^2. \tag{50}$$

We can now write directly the limiting result for the probability distribution $\rho_n(\Delta \vec{w}, \Delta \vec{v})$ after n steps of the random walk, with $n \to \infty$,

$$\rho_n(\Delta \vec{w}, \Delta \vec{v}) = \frac{1}{\sqrt{(2\pi n)^4 \det(\bar{\mathcal{M}})}} \exp \left\{ -\frac{1}{2n} \left[\begin{pmatrix} \Delta \vec{w} & \Delta \vec{v} \end{pmatrix} \bar{\mathcal{M}}^{-1} \begin{pmatrix} \Delta \vec{w} \\ \Delta \vec{v} \end{pmatrix} \right] \right\}. \tag{51}$$

We can use this probability distribution to compute properties of the velocity distribution arising from the fundamental two-particle correlations expressed in $\mathcal{M}$. Thus the expectation value of the velocity at the spatial point $\vec{x}$ is given by

$$\langle \vec{v}(\vec{x}) \rangle = \frac{\int dn \int d\vec{v}(\vec{v}) \rho_n(\vec{x}, \vec{v})}{\int dn \int d\vec{v} \rho_n(\vec{x}, \vec{v})}. \tag{52}$$

As an example, consider a case in which $\langle \delta w_x \delta w_y \rangle = \langle \delta v_x \delta v_y \rangle = 0$, and $\langle \delta w_x \delta v_x \rangle = \langle \delta w_y \delta v_y \rangle = \langle \delta w_y \delta v_x \rangle = 0$, but in which $\langle \delta w_x \delta v_y \rangle \neq 0$. In this case we can write

$$\bar{\mathcal{M}} = \begin{pmatrix} \langle \delta w_x \delta w_x \rangle & 0 & 0 & \langle \delta w_x \delta v_y \rangle \\ 0 & \langle \delta w_y \delta w_y \rangle & 0 & 0 \\ 0 & 0 & \langle \delta v_x \delta v_x \rangle & 0 \\ \langle \delta v_y \delta w_x \rangle & 0 & 0 & \langle \delta v_y \delta v_y \rangle \end{pmatrix}. \tag{53}$$

We can now invert $\mathcal{M}$ and use Eq. (52) to show that

$$\frac{\partial \langle v_y \rangle}{\partial x} = \frac{\langle \delta w_x \delta v_y \rangle}{\langle (\delta w_x)^2 \rangle}, \tag{54}$$

and

$$\frac{\partial \langle v_x \rangle}{\partial x} = \frac{\partial \langle v_{x,y} \rangle}{\partial y} = 0, \tag{55}$$

corresponding to pure shear motion.

In this way, any motion that is linear in the coordinates can be associated with values of the coefficents of the matrix $\mathcal{M}$, corresponding to particular correlations of neighboring grains. This approach has an interesting feature, however, in that it leads to divergent fluctuations in dimensions below $D = 4$.

The generalization of the above formulae to $\rho_n^{(D)}(\vec{x}, \vec{v})$ with $D > 2$ is simple, so we can consider, for instance,

$$\langle (\vec{v}(\vec{x}) - \langle \vec{v}(\vec{x}) \rangle)^2 \rangle = \lim_{N \to \infty} \frac{\int_0^N dn \int d\vec{v}(\vec{v}^2 - \langle \vec{v} \rangle^2) \rho_n^{(D)}(\vec{x}, \vec{v})}{\int_0^N dn \int d\vec{v} \rho_n^{(D)}(\vec{x}, \vec{v})}. \tag{56}$$

Simple power-counting leads immediately to the conclusion that

$$\langle (\vec{v}(\vec{x}) - \langle \vec{v}(\vec{x}) \rangle)^2 \rangle \sim N^{2-D/2}. \tag{57}$$

If we suppose that $\ell \sim \sqrt{N}$ corresponds to some maximum "coherent" length scale that can be probed by the random walk, then we see that

$$\langle (\vec{v}(\vec{x}) - \langle \vec{v}(\vec{x}) \rangle)^2 \rangle \sim \ell^{4-D}. \tag{58}$$

Note that assuming that $I = d/\ell$, as is done in the theory of incline flow of Halsey and Ertaş [4], gives a divergence of the squared velocity fluctuation

$$\langle (\vec{v}(\vec{x}) - \langle \vec{v}(\vec{x}) \rangle)^2 \rangle \sim I^{D-4}, \tag{59}$$

compared to the numerical result

$$\langle (\vec{v}(\vec{x}) - \langle \vec{v}(\vec{x}) \rangle)^2 \rangle \sim I^{-\chi}. \tag{60}$$

The numerical results for the values of χ in $D = 2, 3$ are not definitive on its actual value, although $\chi \approx 1$ in $D = 2$ is likely [6], and a similar and perhaps smaller value seems to hold in $D = 3$ [7].

It is striking that an argument with so little physics predicts a divergence of the velocity fluctuations, and with an exponent similar to that observed numerically [6, 7]. However, the neglect of correlations undermines the quantitative credibility of this argument[1].

Up to now, we have not implemented any requirement that the velocities be determined by the angular motions of the particles. This is seemingly quite straightforward, e.g., the averages of $\delta \vec{v}$ appearing in $\bar{\mathcal{M}}$ can easily be written in terms of the angular motions of the particles, using Eq. (8), for the rolling contacts. Then, as remarked above, the percolation of the contact network insures that we can still construct the master probability distribution for the fluctuations of $\Delta \vec{w}, \Delta \vec{v}$, as in the above argument. There is, however, a subtle and important flaw in this procedure.

Restricting ourselves for the moment to two dimensions, let us try to determine large scale variations in the angular velocities Ω_i using a procedure analogous to that we used above for large scale variations in the velocity. Since the velocity moments for neighboring particles are functions of the sum of the angular velocities, $\vec{v}_i - \vec{v}_j = \hat{z} \times (\vec{w}_i - \vec{w}_j)(\Omega_i + \Omega_j)$, we will restrict ourselves to these variables in computing large-scale variations of Ω, which is feasible provided we consider the combination of two successive steps on the lattice of rolling contacts.

If we construct an analogous formula to Eq. (51) for the evolution of the distribution of Ω over a random walk, we can write

$$\frac{\partial \Omega}{\partial x} = \frac{\langle \delta w_x \delta \Omega \rangle}{\langle (\delta w_x)^2 \rangle}. \tag{61}$$

Note that we have assumed above that there is a stationary distribution $\rho(\theta, \delta \vec{v}_{ij})$. If such a distribution is not a function of position, then there is no local indicator of position arising neither from the angular distribution of contacts, nor from the nearest-neighbor velocity differences. For rolling contacts, this implies, as utilized in the above, a stationary distribution $\rho(\theta, \Omega_i + \Omega_j)$. To use this to evaluate an average of $\Omega_i - \Omega_j$, we can concatenate two subsequent steps in our random walk, from $i \to j \to k$, and write that for this compound step

$$\langle \delta w_{x;i \to k}(\Omega_k - \Omega_i) \rangle$$
$$= \langle (\delta w_{x;i \to j} + \delta w_{j \to k}) \times [(\Omega_k + \Omega_j) - (\Omega_j + \Omega_i)] \rangle \tag{62}$$
$$= d \langle \cos \theta_{ij}(\Omega_i + \Omega_j) - \cos \theta_{jk}(\Omega_j + \Omega_k) \rangle = 0,$$

so that we conclude that it is not possible for a distribution with a stationary $\rho(\theta_{ij}, \Omega_i + \Omega_j)$ to describe a grain packing with any large-scale spatial variation of Ω.

---

[1]Note that a similar argument could be used for a molecular fluid, in which case it would clearly be wrong, by equipartition – highlighting the importance of correlations.

However, inspired by the discussion above regarding the role of $\xi = \frac{d}{2}(\Omega_A - \Omega_B)$ in the honeycomb lattice case, we can immediately find a way around this. Suppose the packing of grains connected by rolling contacts can be described as consisting of two disjoint sub-packings A and B, so that no two A grains, nor any two B grains, are in rolling contact with one another. In this case, we can posit that the distribution used in the random walk depends on whether one is passing from an A grain to a B grain or vice versa, i.e., $\rho_{AB}(\theta, \Omega_A + \Omega_B) \neq \rho_{BA}(\theta, \Omega_B + \Omega_A) = \rho_{AB}(\theta + \pi, \Omega_A + \Omega_B)$, where this latter requirement follows from the reversiblity of the random walk. Then we see that in this case, supposing that i, k are on the A packing, and j is on the B packing,

$$
\begin{aligned}
\langle \delta w_{x; i \to k}(\Omega_k - \Omega_i) \rangle \\
= \langle (\delta w_{x; i \to j} + \delta w_{j \to k})[(\Omega_k + \Omega_j) - (\Omega_j + \Omega_i)] \rangle \qquad (63) \\
= 2d \langle \cos \theta_{ij}(\Omega_i + \Omega_j) \rangle_{AB},
\end{aligned}
$$

where $\langle\rangle_{AB}$ is defined as the integral over $\rho_{AB}(\theta, \Omega_A + \Omega_B)$. Now we can develop linear gradients in Ω on sub-lattice A, provided that the gradient of Ω on the alternating sub-lattice B has the opposite sign, as in the honeycomb lattice solution.

Thus, the state will be characterized by a function $\Omega_A(x, y)$ on the A sub-lattice, and by $\Omega_B(x, y)$ on the B sub-lattice, with

$$
\frac{\partial \langle \Omega_A \rangle}{\partial x} \approx - \frac{\partial \langle \Omega_B \rangle}{\partial x} \approx \frac{\langle \cos \theta_{ij}(\Omega_i + \Omega_j) \rangle_{AB}}{d \langle (\cos \theta_{ij})^2 \rangle_{AB}}. \qquad (64)
$$

This criterion will enforce that relative local particle surface velocities will be small where an A-lattice particle contacts a B-lattice particle, which is suitable to a slowly strained system with frictional contacts. Both Ω_A and Ω_B vary linearly across a shearing region. Thus, the dimensional constraint on the forces, Eq. (48), should still apply for random lattices, resulting in a restriction of the coherently shearing region to a scale $\ell = d/I$. Finally, the overall shear is determined in this case by

$$
\frac{\partial \langle v_y \rangle}{\partial w_x} = \frac{\langle \cos^2 \theta_{ij}(\Omega_i + \Omega_j) \rangle_{AB}}{2d \langle (\cos \theta_{ij})^2 \rangle_{AB}}. \qquad (65)
$$

To build an explicit example of a shearing state on a random graph, suppose that we fix the rotations on the A and B sub-lattices to be simple functions of position alone,

$$
\Omega_A = \Omega_A(\vec{x}) = \Omega_A^{(0)} + \vec{\Omega}_A^{(1)} \cdot \vec{x}, \qquad (66)
$$

$$
\Omega_B = \Omega_B(\vec{x}) = \Omega_B^{(0)} + \vec{\Omega}_B^{(1)} \cdot \vec{x}. \qquad (67)
$$

Then for two particles in contact at positions $\vec{x}$ and $\vec{x} + \delta \vec{w}$,

$$
\Omega_A(\vec{x}) + \Omega_B(\vec{x} + \delta \vec{w}) = (\Omega_A^{(0)} + \Omega_B^{(0)}) + (\vec{\Omega}_A^{(1)} + \vec{\Omega}_B^{(1)}) \cdot \vec{x} + \vec{\Omega}_B^{(1)} \cdot \delta \vec{w}. \qquad (68)
$$

In order for local velocity gradients to be independent of position, as in the above, we need

$$
\vec{\Omega}_B^{(1)} = -\vec{\Omega}_A^{(1)} \equiv \vec{\Gamma}, \qquad (69)
$$

and we also write, for convenience,

$$\Omega_A^{(0)} + \Omega_B^{(0)} = \varphi. \tag{70}$$

Since we have fixed the angular velocities $\Omega_{A,B}$ as functions of position alone, we can suppress the dependence of ρ_{AB} on $\Omega_A + \Omega_B$, and approximate

$$\rho_{AB}(\theta) = \rho_0 + \rho_1 \vec{b} \cdot \delta\vec{w} + O(\delta\vec{w}^2), \tag{71}$$

with, of course,

$$\delta\vec{w} = d(\cos\theta, \sin\theta). \tag{72}$$

Reviewing Eqs. (54, 65), we see that to lowest order in the anisotropy, the numerator of this expression will determine the overall shear structure. Thus let us consider the following tensor,

$$T_{lm,AB} = \langle \delta w_l \delta v_m \rangle_{AB} \approx \int (d\delta\vec{w})(\rho_0 + \rho_1 \vec{b} \cdot \delta\vec{w}) \delta w_l \left[\frac{1}{2}(\varphi + \vec{\Gamma} \cdot \delta\vec{w})\hat{z} \times \delta\vec{w} \right]_m, \tag{73}$$

where we have used the vector structure appropriate to $D = 2$ as well as the fundamental contact rolling Eq. (8). The integral is easily performed, yielding

$$T_{lm,AB} = \frac{d^2}{4}\rho_0\varphi\epsilon_{lm} + \frac{d^2}{16}\rho_1 \left[\vec{b} \cdot \vec{\Gamma}\epsilon_{lm} - b_l(\epsilon_{mp}\Gamma_p) - \Gamma_l(\epsilon_{mp}b_p) \right]. \tag{74}$$

From this formula, it is easy to find, for any $\vec{b}$, the values of φ and $\vec{\Gamma}$ corresponding to any particular shearing motion. Note that it is necessary to have both a linear gradient in the angular velocities (non-zero $\vec{\Gamma}$) and anisotropy in the contact distribution among the two sub-lattices (non-zero $\vec{b}$) in order to generate a shearing motion. Of course, for any such motion, the value of $\vec{b}$ will be determined by the kinetics of the motion, so in practice such motions will be realized as functionals of $(\varphi, \vec{\Gamma})$.

In general, it is not possible to decompose a random lattice into two alternate sub-lattices such that no contacts are created among elements of a single sub-lattice. The situation is analogous to an anti-ferromagnet on a random graph, for which plaquettes bounded by odd numbers of particles are frustrated (see Figure 9) [22]. In order to preserve the structure of the state indicated above, with $\Omega_A \approx -\Omega_B$, we require that enough of the particle contacts are sliding so that all remaining contacts can be rolling contacts between grains on alternate sublattices. It then follows that ρ_{AB} is understood as the distribution across rolling contacts only, not including sliding contacts.

If we wish to apply the reasoning of Eqs. (44–48) to the random lattice, and derive a length scale ℓ beyond which a coherently shearing patch will be destabilized by the disappearance of normal forces, we must specify what we mean by a "coherently shearing patch" in a random lattice. For a general random lattice with some density of "odd" plaquettes, there will be a number of possible choices of which contacts must slide so that the others may roll. From the argument of the preceding paragraph, it is clear that this enumeration problem is exactly analogous

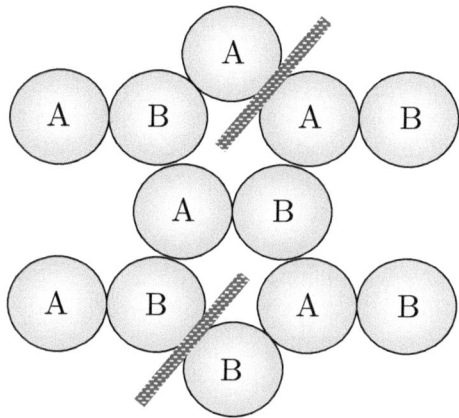

FIGURE 9. In a random lattice, a decomposition of the lattice into two alternating sub-lattices will generally result in "frustrated" contacts, across which sliding must occur.

to the enumeration of the states of an anti-ferromagnet on the corresponding random graph. The statistical mechanics of this latter problem has been studied – in two dimensions, there appears to be a spin-glass phase at zero temperature, which does not extend to finite temperature [21]. Although the granular problem is not a thermal statistical mechanical problem, it is natural to identify the coherence that is destroyed by the disappearing normal forces with the zero-temperature spin glass order of the analogous anti-ferromagnet. With this interpretation of the meaning of "coherence", we again expect

$$\ell = \frac{d}{I} \tag{75}$$

to set the maximum size of a coherent domain, following the argument of Eqs. (44–48). A more dramatic indicator of the existence of the state which we are discussing would be a strong short-ranged anti-ferromagnetic order corresponding to a predominance of rolling contacts.

We can compare our results with numerical simulations reported by the group of Alonso-Marroquín et al. [20]. In a two-dimensional shear cell, this group observed that the macroscopic shearing motions of a dense granular packing (simulating fault gouge) decomposed, on smaller scales, into regions of coherent vorticity (which in our notation corresponds to Δ constant, $\xi \approx 0$), regions of "ball-bearing motion" (corresponding to the counter-rotating motion of our two sub-lattices), and shear zones in which sliding dominated. Due to the predominance of rolling contacts, particularly at high coefficients of friction, the overall macroscopic friction observed in the shear cell was considerably less than the microscopic coefficient

of friction. The decomposition of the solution into regions with qualitatively different properties violates the assumption that there is a uniform distribution $\rho(\theta, \delta\vec{v})$,
on which our argument above was based.

The first of the Alonso-Marroquín motions, corresponding to large-scale vorticity, was not observed in three-dimensional numerical studies specifically de-
signed to look for velocity correlations in chute flows [7]. Also, at high coefficents
of friction, the Alonso-Marroquín kinematics was "earthquake-like", with much of
the slip occuring in discontinuous jumps. We might suspect that such disconti-
uous kinetics, associated with pattern formation, are more characteristic of low
values of I (corresponding to $\ell \gg L$, with L an overall flow scale), than they are
of intermediate values of I.

Finally, we are also able to use this picture to speculate on the nature of the
$\mu \to \infty$ limit. Consider the tangential and normal forces T_c, N_c at a typical sliding
contact, $T_c/N_c = \mu$. These forces will be determined by the overall force balances
subject to the motions of the grains, and we expect that both T_c and N_c will be
$\sim Pd^{D-1}$. Since these forces have a fixed ratio at the sliding contacts, we see that
$\lim_{\mu \to \infty} N_c = 0$, since this will be the solution to the force balances in preference
to a case in which $T_c \to \infty$. Thus, in the limit, the sliding contacts will see their
normal forces driven to zero; i.e., these contacts will be eliminated as physical
contacts. We would thus expect that in the limit $\mu \to \infty$ all of the plaquettes will
have even numbers of sides, and the frustration will be eliminated, in any mobile
state.

4.1. Three-Dimensional Case

The extension of these arguments to three dimensions is straightforward. The
original contact-rolling equations Eqs. (8, 9) clearly apply in three dimensions,
and the diamond lattice plays the same role in $D = 3$ of a potential model system
as did the honeycomb lattice in $D = 2$. To conserve the labor of the author and
the patience of the reader, we are not presenting details of the diamond lattice
kinematics in this work; we only wish to point out that the diamond lattice does
support alternating A and B sublattices analogously to the honeycomb lattice case,
which allows shearing states obeying $\vec{\Omega}_A \approx -\vec{\Omega}_B$ to be constructed.

For random lattices, the arguments of the previous section should hold in
three dimensions just as in two dimensions (indeed, they should be more valid,
since the upper critical dimension of $D = 4$ is closer). Again, the optimal decomposition of the random lattice into two alternating sublattices is analogous to the
problem of determining a ground state for an anti-ferromagnet on a random graph.
Again, odd plaquettes must have at least one sliding contact, corresponding to a
frustrated bond in the random anti-ferromagnet. In three dimensions, these odd
plaquettes can be viewed as threaded by "odd lines", introduced by Rivier and
Duffy [23]. For pure shear, we expect the direction of the typical angular velocity
to be perpendicular to the shearing motions, so that the analogy is to an Ising

anti-ferromagnet, and not to a Heisenberg anti-ferromagnet. Note that the degenerate case in which particles rotate about an axis through the contact between the particles is assumed not to play a significant role.

5. Conclusions

The principal conclusions of this article are:

1. Frictional packings, dominated by rolling contacts, cannot be mobile above a coordination number $Z_c = 3$ ($D = 2$) or $Z_c = 4$ ($D = 3$). These thresholds are consistent with those obtained using similar arguments by authors studying the isostaticity of static packings. These criteria are modified in a straightforward manner if some of the contacts are sliding as opposed to rolling.

2. The honeycomb lattice in $D = 2$ offers a case in which the rolling kinematics can be exactly solved. The most surprising feature of the result is that the scale of the angular velocities ω grows linearly with the size of the system, although the average of the angular velocity is moderated by the fact that the rotations on two alternating sub-lattices roughly cancel.

3. The dynamics of the honeycomb lattice can also be solved. The conclusion is that the requirement that all normal forces be compressive can only be satisfied for packings smaller than ℓ, with

$$\ell \sim d/I, \tag{76}$$

where I is the Inertia Number, given by Eq. (3).

4. For random lattices, a Fokker-Planck approximation to the kinetics yields the same key result as for the honeycomb lattice, i.e., that the typical angular velocity grows linearly with the size of a coherently rolling region. Again, the average angular velocity is much smaller. The presence of odd plaquettes of particles in contact will require compensating sliding contacts even if the coefficient of friction is large; the statistics of these sliding contacts are analogous to those of frustrated bonds in random anti-ferromagnets.

There remains the very interesting question of the nature of the flow for scales larger than ℓ. We can understand the role of ℓ by considering the free volume in the system. For flows in systems of size $L < \ell$, while free volume might be created by collisions generating a granular temperature, such free volume can be removed from the system through the role of inelasticity (or friction) in quenching the granular temperature. On the other hand, for $L > \ell$, free volume must be created and persist in the system, regardless of the quenching effect of inelasticity. It is simply no longer possible for all of the stress-transmitting contacts to have lifetimes $> \dot{\gamma}^{-1}$, consistent with observations of the relatively short duration of most two-particle collisions [13]. Although we have focused on the continuous motion of systems for which $L < \ell$, the fact remains that the chute flow phase diagram, or the results of Alonso-Marroquín mentioned above, emphasize the difficulty of flowing granular systems of size $< \ell$, suggesting that as a practical matter free volume is necessary to achieve a robust flowing state.

I would claim that the rheology of this state, in which free volume plays a significant role, while many neighboring particles also have rotational motions effectively locked to one another, is probably poorly addressed by conventional or modified kinetic theory approaches, due to the presence of long-range correlations in particle motion. The interpretation of the GDR MiDi or Pouliquen flow rule type rheologies in terms of a length scale ℓ obviously disallows such approaches. In this article, I have shown that this length scale ℓ does have a clear physical meaning for $L < \ell$; however, it remains to be seen what aspects of this meaning persist for systems of size $L > \ell$.

Acknowledgment

I am grateful to all of my collaborators on the work summarized in this article, including O. Baran, T. Börzsönyi, G.S. Grest, J. Lechman, D. Levine, S. Plimpton, L. Silbert, and especially D. Ertaş. D. Ertaş has also shared a number of results prior to publication. I would also like to thank R. Blumenfeld for bringing the existence of Ref. [22] to my attention.

References

[1] A.N. Schofield and C.P. Wroth, *Critical State Soil Mechanics,* (McGraw-Hill, 1968).

[2] C.K.K. Lun and S.B. Savage, J. Appl. Mech. **54** (1987), 47.

[3] V. Kumaran, J. Fluid Mech **599** (2008), 121; and references therein.

[4] D. Ertaş and T.C. Halsey, Europhys. Letts. **60** (2002), 931; T.C. Halsey and D. Ertaş, arXiv: cond-mat/0506170.

[5] J.T. Jenkins, Phys. Fluids **18** (2006), 103307.

[6] G.D.R. MiDi, Eur. Phys. Journ. E **14** (2004), 341.

[7] O. Baran et al., Phys. Rev. E **74** (2006), 051302.

[8] O. Pouliquen, Phys. of Fluids **11** (1999), 542.

[9] L.E. Silbert et al., Phys. Rev. E **64** (2001), 051302–1.

[10] T. Börzsönyi, T.C. Halsey and R.E. Ecke, Phys. Rev. E **78** (2008), 011306.

[11] R.A. Bagnold, Proc. Roy. Soc. London A **225** (1954), 49; **295** (1966), 219.

[12] L.E. Silbert, J.W. Landry, and G.S. Grest, Phys. Fluids **15** (2003), 1.

[13] L.E. Silbert, G.S. Grest, R. Brewster, and A.J. Levine, Phys. Rev. Lett. **99** (2007), 068002.

[14] S.F. Edwards, Physica A **249** (1998), 226; S. Alexander, Phys. Rep. **296** (1998), 65.

[15] T.C. Halsey, Phys. Rev. E **80** (2009), 011303.

[16] L. Rothenburg and N.P. Kruyt, Int. J. Solids and Struct., **41** (2004), 5763.

[17] The coefficient of the sliding contact term is slightly different from the force-counting argument in L.E. Silbert, D. Ertaş, G.S. Grest, T.C. Halsey, D. Levine, Phys. Rev. E **65** (2002), 051307, due to the fact that sliding constrains only one kinematical variable (the distance of the centers of the particles), while the force counting at a sliding contact is determined by the criterion of Coulomb yield.

[18] The criterion that a contact be sliding reduces the number of kinematical constraints, but introduces an equivalent number of Coulomb constraints linking T to N, as in T.C. Halsey and D. Ertaş, Phys. Rev. Lett. **83** (1999), 5007.

[19] Traction and dissipation at a rolling contact are treated in K.L. Johnson, *Contact Mechanics*, (Cambridge Univ. Press, Cambridge, 1985) p. 306 ff.

[20] F. Alonso-Marroquín et al., Phys. Rev. E **74** (2006), 031306.

[21] M. Weigel and D. Johnston, Phys. Rev. B **76** (2007), 054408; for a review see R. Moessner, Can. J. Phys. **79** (2001), 1283.

[22] A similar approach, also based on frustration, was used by R.C. Ball and R. Blumenfeld to approach static states and perturbations thereof in R.C. Ball and R. Blumenfeld, Phys. Rev. Lett. **88** (2002), 115505–1; R. Blumenfeld, Physica A **336** (2004), 361.

[23] N. Rivier and D.M. Duffy, J. Phys. C: Solid State Phys. **15** (1982), 2867.

Thomas C. Halsey
ExxonMobil Upstream Research Company
3120 Buffalo Speedway
Houston, TX 77098 USA
e-mail: **thomas.c.halsey@exxonmobil.com**

Glasses and Grains, 137–157
© 2011 Springer Basel AG

Grains, Glasses and Jamming

Olivier Dauchot

Abstract. On the one hand, very strong similarities have been reported between the dynamical behaviour of vibrated dense packings of grains and supercooled liquids at the onset of the glass transition. On the other hand a lot of attention has been paid to the mechanical properties of granular packings close to jamming. Here, after a necessary clarification of the different meanings of jamming, we propose to review some of the hereabove mentioned similarities. We discuss, through the comparative study of two different granular systems and one super-cooled repulsive liquid, how elementary relaxation events spatio-temporally organize close to the glass transition. The decomposition of the microscopic dynamics into a "trivial" vibration-like motion and rare "cage jumps", which dominate the relaxation dynamics, is a common feature of these systems. As a result the origin of the vibrating motion, thermal or mechanical, becomes irrelevant, hence the similarities observed at the macroscopic scale. We discuss the evolution of the spatio-temporal organisation of these relaxation events when approaching the glass transition and investigate its relation to softness of the structure.

1. Introduction and preliminary remarks

Everyday life tells us that matter acquires rigidity when it cools down – lava flows turn into solid rocks – or when it is compacted – remember kids packing down the sand to build up castles on the sea shore. As suggested by these examples, this is not only the case for materials, that crystallize at low temperature or high pressure. It also happens for disordered media such as foams, emulsions, colloidal suspensions, granular media and glasses, which can jam in a rigid disordered state [40, 53].

When a liquid is cooled down fast enough to avoid crystallization, it enters the metastable super-cooled regime. Further decreasing the temperature, one observes a dramatic slowing down of the dynamics and a corresponding increase of the viscosity over several decades. Eventually, depending on the cooling rate, the system falls off equilibrium, ages, and becomes an amorphous rigid material: this is the glass transition. Another well-studied instance of glassy behaviour is that of hard spheres at thermal equilibrium [47]. In this case the transition is controlled by the packing fraction and the colloidal suspensions are a good experimental model of this idealized system as clearly illustrated in the chapter by D. Weitz in the present book.

In the last two decades an intense activity in the field of granular matter has reported strong experimental evidence of such glassy dynamics in dense granular media under low mechanical excitations: dynamical slowing down, aging phenomena, rigid response to external stress [20]. Here also the control parameter of the dynamics is the packing fraction. These observations suggest a possible unified description despite the very essential difference between grains and supercooled liquids: for the former the dynamics is dissipative and the system is forced into a steady state by various types of mechanical excitations, whereas for the latter the dynamics is that of thermodynamic equilibrium.

In 1998 Liu and Nagel [40] presented their provocative jamming phase diagram, while discussing the concept of fragile matter introduced by Cates *et al.* [16] in the context of granular materials. The purpose of this (Temperature-Stress-Density) diagram was to generalize the discussion both to the case of microscopic systems with attractive interactions, which unjam as one raises the temperature, and to the case of stressed macroscopic systems with repulsive interactions, which unjam as one applies an incompatible stress. According to this diagram, a glass would have a lower glass transition temperature under high shear stress. Likewise, a jammed granular material or foam would have a lower yield stress when random motions are present. One must realize the boldness of such a proposition, which encompasses in a unique framework the glass transition [3, 22] and the emergence of a yield stress [4], two of the most challenging issues in modern condensed matter physics.

As long as jamming is understood in the loose sense described above, the jamming diagram is a challenging proposal and as such it has been a source of inspiration for many fascinating studies. However, jamming also has a very precise meaning when it refers to a zero temperature and zero stress transition which can be defined in purely geometric terms and is closely related to the notion of random close packing [44, 45]. This transition occurs when a given packing of particles can not be compressed further without allowing overlaps between particles. Because this last transition does not need to invoke thermal equilibrium, it rapidly became a favorite candidate for explaining the glassy behaviour of granular media first and, by virtue of the 'universality' of the jamming diagram, of thermal systems also. This line of thought has been reinforced by the existence of remarkable scaling properties observed on the approach to 'point J' for various kinds of repulsive particles [25, 28, 39, 46]. And naturally this led to the conjecture that the properties of this 'critical point' can influence the physical behaviour of dense particle systems throughout the whole jamming diagram.

Confusion has been endemic until recently, when strong theoretical [47] and numerical evidence [6, 7, 18] has been provided for the ideal case of frictionless hard spheres. First, even in the thermodynamic limit, jamming transitions occur along a continuous range of volume fractions, contrasting with the idea of a unique 'point J'. Second it was shown that at the glass transition, where the relaxation time becomes infinite, pressure remains finite. On the contrary, at the jamming transition of hard spheres, pressure must diverge because of the infinitely hard repulsive interaction. More precisely the structure – the averaged positions of the particles – is frozen at the glass transition, but there is still room for vibrations around this configuration. Jamming occurs at higher packing fractions when there is no more room to accommodate the increase of the packing fraction without overlapping the particles. Clearly the jamming and the glass transitions are distinct, the former happening in the glass phase. As a matter of fact the same holds in the case of crystallisation. The relaxation time of the liquid diverges, but the pressure remains finite. One can still increase the packing fraction of the crystal until it reaches an ordered close packing: only then is the crystal jammed! In the following we shall concentrate on the glass transition of dense granular media and super-cooled liquids. Accordingly the word jamming here should be understood in its most general and common sense, and not as referring to any specific transition.

Now that the above necessary clarifications have been made, we can turn ourselves to the main purpose of the present notes. We first present a brief review of the glassy phenomenology in granular media. We then illustrate, through the comparative study of two different granular systems and one super-cooled repulsive liquid, how elementary relaxation events spatio-temporally organize and develop dynamical heterogeneities. In the case of one of the granular experiments, we further discuss how this spatio-temporal organization evolves when approaching the glass transition. In the case of the super-cooled liquid simulations, we can investigate the relation of these dynamical properties to the structural properties of the material. Finally in a last section, we discuss what appears to us as some of the most challenging issues for the near future.

2. Glassy behaviour of vibrated grains

In this part we will briefly present a selection of experimental results, which underlines the similarity between granular media and super-cooled liquids close to the glass transition. It is assumed that the reader is familiar with the glass transition. He might otherwise refer to the other chapters of the present book. A more detailed review of these similarities can also be found in [20]. We first review macroscopic behaviours observed under compaction and then concentrate on the microscopic dynamics at the origin of these striking similarities.

2.1. Macroscopic evidence

Generically, one considers a three-dimensional sample of grains under compaction. The packing is prepared in a reproducible low density initial stage and is then vibrated with an amplitude a and a frequency ω. Different shapes of vibration can be applied, but the overall control parameter is usually defined as $\Gamma = a\omega^2/g$.

The column density is then monitored for instance with capacitors [36], or via X-ray absorption measurements [49, 48]. Other ways of compacting include cyclic shear [42] or flow pulses in a fluidized bed [50].

2.1.1. Slow relaxations. The very first evidence of a "glassy" behaviour in dense granular media under compaction is the very slow relaxation towards a stationary state with a well-defined volume fraction. Figure 1 presents the dynamics of compaction as observed in Chicago [36] and Rennes [49, 48]. In both cases, the logarithmic time scale emphasises the very slow dynamics and in the Chicago experiment (Figure 1(a) it is not even clear that a stationary state is reached within the duration of the experiment. In the case of the experiment in Rennes (Figure 1(b), a stationary state is obtained, but for large vibration amplitudes only. For small values of Γ, it is difficult, if not experimentally impossible, to reach a steady-state by merely applying a sufficiently large number of taps of identical intensity. Nowak *et al.* [43] showed that, in this case, it is possible to reach a steady state by annealing the system 1(c). Experimentally, the value of Γ is slowly raised from 0 to a value beyond $\Gamma^* \simeq 3$, above which subsequent increases as well as decreases in Γ at a sufficiently slow rate $d\Gamma/dt$ lead to reversible, steady-state behaviour. If Γ is rapidly reduced to 0 then the system falls out of the steady state branch. Along the reversible branch, the density is monotonically related to the acceleration. As Γ is increased both the magnitude of the fluctuations around the steady state and the amount of high-frequency noise increase.

Various fits have been proposed to describe these experimental data among which is the one introduced by Kohlrausch [38], Williams and Watts [32], often denoted as the KWW law, which is commonly observed in the relaxation of thermal glasses. Also, in the "Rennes" experiment, the relaxation time dependence is reminiscent of an Arrhenius law $\tau = \exp(\Gamma_0/\Gamma)$ for $\Gamma > 1$, suggesting the existence of activated processes (Figure 1d). For $\Gamma < 1$, one clearly observes a sharp increase of the relaxation times. The slope variation in the log-lin plot, which indicates a jump in the 'energy barrier' of the mechanically activated processes suggested by the Arrhenius laws, finds a natural interpretation in the difference of energy landscape seen by a grain, whether it lifts off or not!

2.1.2. Aging. More specific behaviours of glasses such as aging and memory effects have also been observed in granular media under compaction. Using multi-speckle diffusive wave spectroscopy (MSDWS) to probe the micron-scale dynamics of a water saturated granular pile submitted to discrete gentle taps, Kabla and Debregeas [34] experimentally demonstrated aging effects. The pile is first prepared in a reproducible way at low volume fraction, then submitted to high amplitude taps until it reaches a prescribed packing fraction. Only then is the dynamics of contacts probed by submitting the cell to very gentle taps. Figure 2(a) displays the compaction curves during the full procedure. One recognises a typical compaction curve during the first stage. In contrast, the low intensity vibrations do not induce significant further evolution of the packing fraction except for initially very loose packs. In order to quantify the internal dynamics, one measures the

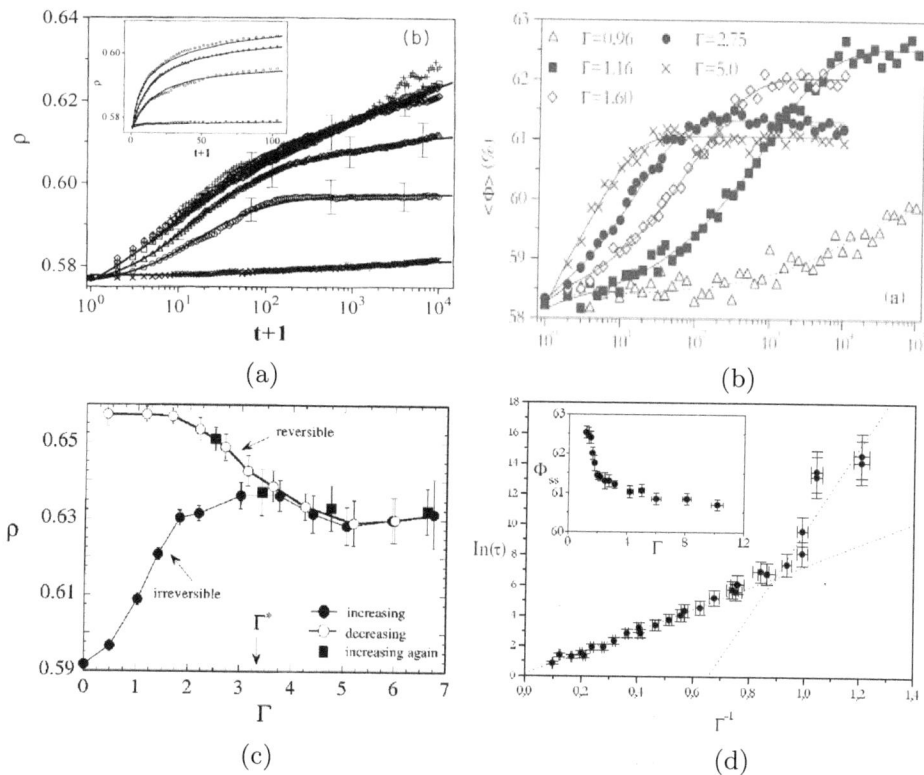

FIGURE 1. Compaction experiments. (a): "Chicago" experiment [36]: packing density ρ as a function of the logarithm of the number of taps for various amplitudes of vibration ranging from $\Gamma = 1.4$ to 5.4 (inset is the same plot in linear scale); (b): "Rennes" experiment [49, 48]: temporal evolution of the mean volume fraction for different tapping intensities ranging from $\Gamma = 0.96$ to 5.0; (c): Steady state branch from [36]: the sample is prepared in a low density initial configuration and then the acceleration amplitude is first slowly increased – solid symbols – and then decreased – open symbols. – The upper branch is reversible, see square symbols. (d): Dependence of the relaxation time as a function of the vibration amplitude in "Rennes" experiment. Inset: variation of the final volume fraction in the cases where a steady state is actually reached.

intensity correlation $g(t_w, t)$ of speckle images – produced by the multiple scattering of photons through the sample –, taken between taps at time t_w and $t_w + t$, where t_w is the elapsed number of taps since the beginning of the gentle vibration. Figure 2(b) shows three correlation functions obtained with the same sandpile at

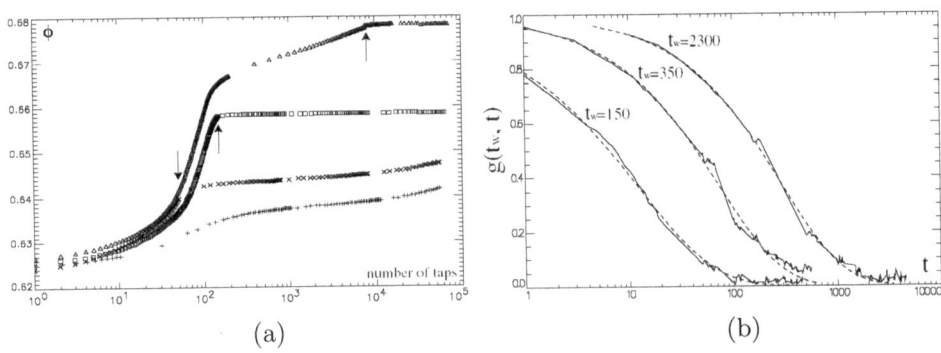

FIGURE 2. Aging in a gently vibrated ($\Gamma \simeq 1$) granular media [34] (a): Evolution of the packing fraction for four experimental runs. Each run consists of a first step in which high amplitude taps allow rapid compaction of the sample, followed by a sequence of gentle vibrations, during which the internal dynamics is probed. The arrows indicate the change in tapping intensity. (b): Two-time relaxation curves for different waiting time.

different values of t_w. These functions, well fitted by stretched exponentials, clearly demonstrate an increase of the relaxation time with t_w. This dynamical arrest is the signature of the aging behaviour as exhibited in various glassy systems.

Altogether, we have seen in this section that the jamming of granular media shares strong similarities – exceedingly slow relaxation and aging – with the glassy dynamics of super-cooled liquids. These similarities are not trivial given the very distinct microscopic processes underlying the dynamics in both systems. In glassy liquids, relaxation occurs by thermally activated rearrangements of the structure. In granular materials, the thermal environment is ineffective and relaxation results from the local yielding of contacts triggered by externally applied vibrations. Still the phenomenologies recorded at the macroscopic level suggest the existence of similar mechanisms at the "microscopic" scale, that is the molecular scale for structural glasses and the grain scale for granular media.

2.2. Microscopic Dynamics: Granular media vs Structural Glasses

At the end of the day, forgetting about thermal equilibrium vs. mechanical steady state issues, the macroscopic properties must find their origin in the complex interplay of the trajectories of the individual particles, whether they are grains or molecules. Bidimensional experiments in granular media, confocal microscopy for the study of colloids and numerical simulations give access to the instantaneous positions of many particles, making available formidable data sets to investigate the dynamics at the microscopic scale. We will now review some of the recent results obtained in these systems at the "microscopic" scale.

2.2.1. Systems. Most of the results described below come from the analysis of three sets of data stemming from two different granular media experiments, a Cyclic Shear Experiment (CSE) [41, 13] and a Fluidized Bed Experiment (FBE) [1, 35, 14], and one Soft Repulsive Particle simulation (SRPS) [15]. A detailed description of these experiments and simulations can be found in the original papers. Here are some of their essential characteristics:

- A picture of the CSE can be found in the chapter by G. Biroli in the present book. A bi-dimensional, bi-disperse granular material, composed of about 8,000 metallic cylinders of diameters 5 and 6 mm in equal proportions, is sheared quasi-statically in a horizontal deformable parallelogram. The shear is periodic, with an amplitude $\theta_{max} = \pm 5°$. The packing fraction is maintained constant ($\phi \simeq 0.84$). Images are taken at each cycle.
- The bi-dimensional fluidized bed of beads is made of a 1:1 bidisperse mixture of steel beads of diameters 3.18 and 3.97 mm, confined to a circular cell. Bead motion is excited by an upward flow of air at a fixed superficial flow speed of $545 \pm 10 \text{cm.s}^{-1}$ (resp. $500 \pm 10 \text{cm.s}^{-1}$) for the three loosest (resp. densest) packing fractions. We will consider packing fraction ranging from $\phi = 0.758$ to $\phi = 0.802$.
- The model of supercooled liquid is a bi-dimensional non-additive binary mixture of $N = 5,760$ particles enclosed in a square box with periodic boundary conditions, interacting via purely repulsive potentials of the form $u_{ab}(r) = \varepsilon(\sigma_{ab}/r)^{12}$. The mole fraction of the smaller particles is taken to be $x_1 = 0.3167$. Molecular dynamics simulations were carried out at constant NVT (T=0.4) using the Nosé-Poincaré Hamiltonian [11] after equilibration at constant NPT as described in [59].

In all cases, the length scale unit is the typical size of the particles and the time unit is chosen in such a way that the structural relaxation time τ_α, defined as the time required for the self-intermediate scattering function to decay of $1/2$, equals 10^3.

2.2.2. Trajectories, cage-jumps and statistics of the displacements. Typical trajectories $r_p(t)$ for the three systems considered here are shown on Figure 3. They exhibit remarkable features which are common to both the granular systems and the repulsive liquid. The particles perform localized, vibrational motion around a metastable position, as in a disordered solid, interrupted by quasi-instantaneous "cage jumps". Note that the size of the jumps is distributed, and represents on average only a small fraction of the particle size, implying that jumps probably result from cooperative events involving a large number of particles moving by a small amount. As a matter of fact, in all these systems the packing fraction is rather large and local re-arrangements necessarily lead to displacements of neighboring particles – this observation is at the root of the idea of cooperative motion and dynamical heterogeneities.

This very specific dynamics immediately translates into the statistics of the displacements. Recently Chauduri *et al.* [17] compared the distributions of the

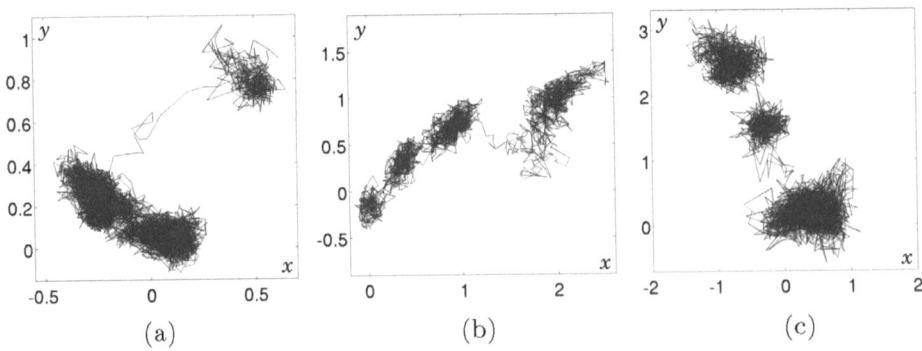

FIGURE 3. Trajectories of one particle (a): in a cyclic shear granular experiment over 5,000 time steps; (b): in the fluidized bed granular experiment over 3,000 time steps; (c): in the soft repulsive liquid simulation over 4,000 time steps.

displacements performed by the particle during a time interval τ for different systems, namely two super-cooled liquids, a colloidal suspension and the CSE granular experiment(see left side of Figure 4). These distributions have the same structure over a broad time window comprising the structural relaxation. They are all made of a central gaussian part corresponding to the short time vibration and large exponential tails associated with the rare and important displacements: the "cage jumps". The width of these distributions also reflects the existence of the cages: the mean-square displacements plotted on the right side of Figure 4 for other but similar systems exhibit a subdiffusive plateau at intermediate timescales. This non-Fickian character of single particle displacements in systems exhibiting glassy dynamics is well known and it can be associated with the non-exponential decay of the time correlation functions. Again one observes the very strong similarities among the very different systems considered here: Lennard-Jones and Hard spheres liquids, colloidal suspensions, granular media.

As far as the comparison between granular media and super-cooled liquids is concerned, the important observation is that the vibrational part of the motion separates from the "cage jumps" dynamics. Following this line of thought, it was suggested to isolate those cage jumps using for instance the iterative algorithm introduced in [13] and to evaluate their relative importance in the structural relaxation. The local relaxation between time t and $t + \tau$ is quantified by:

$$Q_{p,t}(\tau) = \exp\left(-\frac{||\Delta \vec{r}_p(t, t+\tau)||^2}{2\sigma(\tau)^2}\right), \tag{1}$$

where $\Delta \vec{r}_p(t, t+\tau) = \vec{r}_p(t+\tau) - \vec{r}_p(t)$ is the displacement of the particle p, between t and $t + \tau$ and $\sigma(\tau)^2 = \langle ||\Delta \vec{r}_p(t, t + \tau)||^2 \rangle$ is the root mean-square displacement on a lag τ. The interpretation is straightforward: when a particle p moves less,

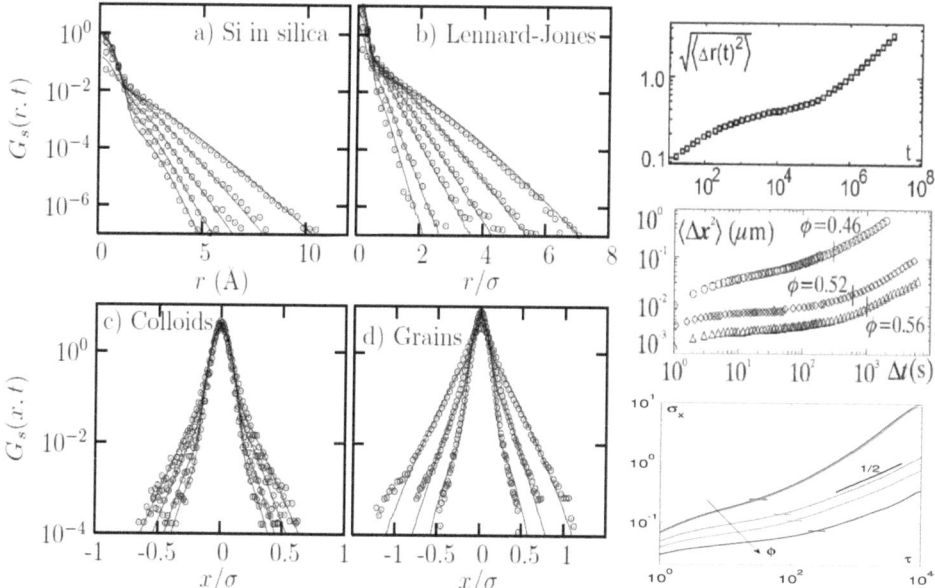

FIGURE 4. Left: Probability distribution of the particles displacements during a lag time τ for silicon atoms in silica, Lennard-Jones particles, hard sphere colloids and grains (from [17]). The data are fitted with a stochastic model of the cage jumps events (see [17]). They exhibit a Gaussian central part and a fat, exponential tail. (a) and (b) show the distributions of $|r_p(t+\tau) - r_p(t)|$, (c) and (d) the distributions of $x_p(t+\tau) - x_p(t)$. Right: Root mean-square displacements as a function of τ for (top) polydisperse hard spheres from [23], (middle) colloids from [56] and (bottom) granular media from the fluidized bed experiment [1, 14].

respectively more, than $\sigma(\tau)$ between t and $t+\tau$, $Q_{p,t}(\tau)$ remains close to one, respectively decreases to zero. Averaging this quantity over all particles, one obtains $Q_t(\tau) = \langle Q_{p,t}(\tau)\rangle_p$, which evaluates the overall relaxation of the system between t and $t+\tau$. Typically the relaxation time τ_α is then given by $\langle Q_t(\tau_\alpha)\rangle_t = 1/2$. As we shall see below, $Q_t(\tau)$ is a highly fluctuating quantity and one can identify a timescale τ^*, for which the fluctuations of $Q_t(\tau)$ are maximal. Figure 5 compares the relative $Q_t(\tau^*)/\langle Q_t(\tau^*)\rangle_t$ with $P_t(\tau^*)/\langle P_t(\tau^*)\rangle$, the relative percentage of particles that have not jumped during the same lag τ^*. The correspondence is excellent: the bursts of cage jumps are responsible for the major relaxation events of the system.

At this point, we have the evidence that the individual particle dynamics in the various systems considered here, ranging from thermal super-cooled liquids to colloids and athermal granular media, are very similar and decompose into a

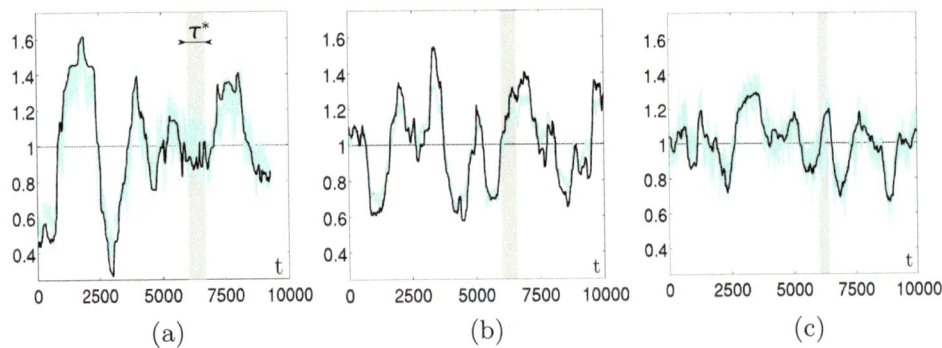

FIGURE 5. Cage jumps and structural relaxation: comparison between the relative averaged relaxation $Q_t(\tau^*)/\langle Q_t(\tau^*)\rangle_t$ (in cyan) and the relative percentage $P_t(\tau^*)/\langle P_t(\tau^*)\rangle_t$ of particles that haven't jumped between t and $t+\tau^*$ (in dark), for (a) the CSE ($\tau^* = 720$), (b) the FBE at $\phi = 0.773$ ($\tau^* = 611$) and (c) the SRPS ($\tau^* = 428$).

somewhat trivial jerky dynamics around a metastable position interrupted by the cage jumps, responsible for the relaxation of the structure. This very important result is at the root of the similarities reported among the glassy dynamics of these very different systems. Since the relaxation is fully encoded in the cage jumps, the vibrational part of the dynamics – whether it has a thermal or a mechanical origin – averages out, and the details of it do not contribute to the glassy dynamics. We now take advantage of these similarities to describe in the same framework the collective dynamics responsible for the glassy dynamics in all these systems.

3. Dynamical heterogeneities in glassy systems

In particular we want to describe the so-called dynamical heterogeneities, which have recently attracted a lot of attention [31, 26, 5, 9]. In this section we will see how dynamical heterogeneities emerge from a hierarchical spatio-temporal organisation of the cage jumps. Then, we shall see that in the case of granular media at least, this spatio-temporal organisation evolves in favour of less facilitation when approaching the glass transition. Finally, we will take advantage of the numerical simulations of the repulsive liquids to discuss the relation of the dynamics with the underlying structural properties.

3.1. Spatio-temporal organisation of the cage jumps

Dynamical heterogeneities are a key characteristic of glassy dynamics in thermal systems [26, 31, 5, 33, 37] colloidal suspensions [55, 19] and granular media [21, 35, 13]. They were first proposed to explain the stretched exponential relaxation of super-cooled liquids. At low enough temperature or high enough packing fraction, the dynamics becomes heterogeneous: domains of slow and fast relaxation coexist in real space and slowly evolve on long time scales. The length-scale associated with these heterogeneities has been argued to be at the origin of the quasi-universal behaviour of glassy systems. In particular, it suggests that the slowing down of the dynamics is related to a collective phenomenon, possibly to a true phase transition. Many different possible origins of these heterogeneities have been highlighted in the literature: dynamic facilitation [30], soft modes [12, 59], proximity to a mode coupling transition [24, 10], growing amorphous order [9], etc. Providing a microscopic explanation for this phenomenology has become a central issue.

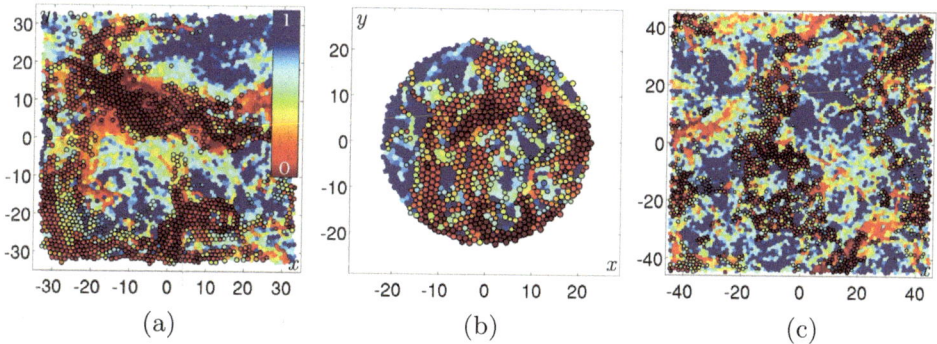

FIGURE 6. Spatial field of the local relaxation $Q_{p,t}(\tau^*)$, for (a) the CSE ($\tau^* = 720$), (b) the FBE at $\phi = 0.773$ ($\tau^* = 611$) and (c) the SRPS ($\tau^* = 428$). Particles jumping between t and $t + \tau^*$ are represented with black circles, and lie preferentially in the moving areas.

It was briefly mentioned in the previous section that the fluctuations ob-served in the temporal evolution of the spatially averaged $Q_t(\tau) = \langle Q_{p,t}(\tau) \rangle_p$ (see Figure 5) are maximal for a given timescale τ^*. These fluctuations can easily be understood as the result of the existence of regions of correlated particles sharing the same relaxation dynamics, exactly like Ising domains being responsible of the fluctuations of the magnetization in a system of spins. Hence τ^* corresponds to the timescale at which these domains are the largest: on shorter timescales the particles move on small enough distances and they don't feel their neighbors; on larger timescales the domains form and deform and the spatial correlation dies away. In practice, τ^* and τ_α are of the same order.

The simplest way to realize the importance of the dynamical heterogeneities is to visualize the spatial field of the local relaxation $Q_{p,t}(\tau)$. Figure 6 displays the fields $Q_{p,t}(\tau^*)$, for the two granular experiments (CSE and FBE) and the repulsive liquid simulation (SRPS), together with, superposed on top of it, the location of the cage jumps, that have occurred between t and $t + \tau^*$. Again the similarity among the three different systems is amazing. There are some differences, as also reflected by the temporal evolutions of $Q_t(\tau^*)$ on Figure 5: clearly the domains are larger in the CSE than in the FBE, and in the FBE than in the SRPS. However, one must remember that these systems are a priori not at the same distance from their glass transition and that the domains grow in sizes when approaching the transition. This last effect is usually quantified by computing the so-called dynamical susceptibility $\chi_4^* = N\left(\langle Q_t(\tau^*)^2\rangle - \langle Q_t(\tau^*)\rangle^2\right)$, exactly in the same way as the magnetic susceptibility finds its origin in the spatial correlation of the local magnetization.

Also shown in Figure 6 are the cage jumps responsible for the relaxation of the dynamics. The distribution in space and time of these events is far from homogeneous. The left panel of Figure 7 illustrates how cage jumps form clusters in space, occurring on a relatively short time scale $\tau_{cluster}$. The distribution of the lag times separating two adjacent clusters can be described by the superposition of two distributions: one for the long times corresponding to the distribution of the time spent by the particles in its cage τ_{cage}, and one for the short delays between adjacent clusters τ_{corr}. When the ratio of these two timescales is large enough, the clusters form well-separated avalanches. Selecting a time interval of duration τ^* initiated at the beginning of a given avalanche, the middle panel of Figure 7 displays the spatial organization of the clusters in the avalanche. One can see how the clusters spread and build up a region of identical temporal decorrelation. This mechanism is a perfect illustration of facilitation: a local relaxation has a very high probability of happening nearby another relaxation after a certain time, which is short compared to the macroscopic relaxation time but large compared to the microscopic one. In the present case, the avalanches have a finite duration: they *are* the dynamical heterogeneities. We shall see in the next section that the scenario depends on the distance to the glass transition.

Finally, the most remarkable fact is that the spatio-temporal organization of the cage jumps is also remarkably similar in both types of system, as demonstrated by the values provided in the table on the right of Figure 7, where the timescales have been rescaled with the relaxation timescale arbitrarily set to $\tau_\alpha = 1000$. This observation suggests the possibility of universal mechanisms, which could be captured at the level of coarse-grained models, such as kinetically constrained models (KCM). However, we shall see now that the recent knowledge acquired from the study of the cage jumps organisation also draws new constraints on these models and calls for alternative proposals.

3.2. Towards the glass transition

We just saw the important role played by dynamical facilitation. Effective models based on kinetic constrains [29, 30] posit that this effect is the underlying cause of particle mobility by assuming that a region of frozen atoms can recover mobility only when it is adjacent to a region already mobile. Within the models this is due to the existence of mobility inducing defects, which cannot disappear (or appear) except if there is another defect nearby. This constraint implies that local relaxations cannot start or end without correspondingly being preceded or followed in space and time by other local relaxations. We will refer to this property as

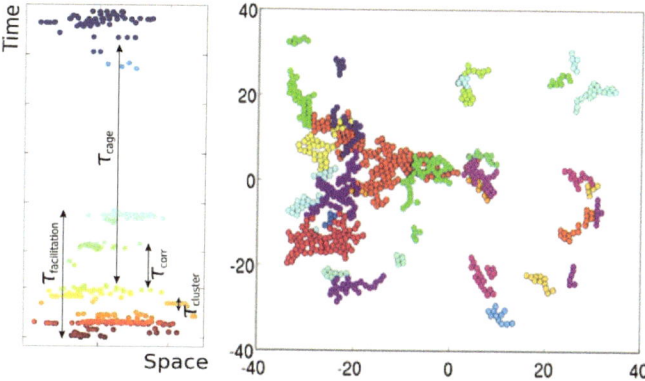

	CSE	SRPS
τ_α	1000	1000
τ^*	915	1078
τ_{corr}	155	240
τ_{cage}	1348	1746
$\sigma(\tau^*)$	0.12	0.29
ξ_4	3.1	2.9

FIGURE 7. Spatio-temporal organization of the cage jumps: Left: Sketch of the spatio temporal organization of the cage jumps. Center: Spatial location of cage jumps, showing how cage jumps facilitate each other to form dynamical heterogeneities in the CSE (from [13]. Right: Comparison of the time and length scales involved in the spatio-temporal dynamics for the CSE and the SRPS.

conservation of dynamical facilitation. In other approaches [8], instead, dynamical facilitation is an important piece of the theoretical description but not the driving mechanism of glassy dynamics.

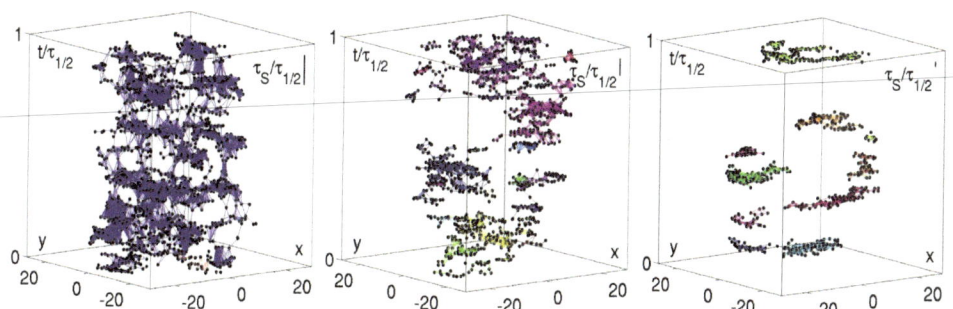

FIGURE 8. Facilitation patterns in space and time during the typical relaxation time $\tau_\alpha(\phi)$ for 3 packing fractions: from left to right $\phi = 0.780, 0.791, 0.802$ (from [14]). The two directions of space are in the horizontal plane and time is the vertical axis. The ratio $\tau_S/\tau_{1/2}$ is given in the upper-right corners. Jumps are represented with black dots, and all possible tetrahedrons which edges are the facilitating links between jumps are shown, forming volumes. Each separate connected structure has a different color.

The dynamical process leading to the avalanches reported in the previous section provides a clear evidence of the important role played by dynamical facilitation: a local relaxation due to a cluster of cage jumps is typically followed nearby in space and in time by another cluster relaxation, and so on and so forth until the entire avalanche process is formed in a typical facilitation timescale τ_f. However, the existence of finite duration avalanches already clearly demonstrate that, in this particular case, dynamical facilitation is not conserved. In order to understand precisely the role played by dynamical facilitation, we will now characterize the evolution of the avalanche process when approaching the glass transition, focusing on the case of the FBE.

The way in which clusters aggregate and the resulting facilitation patterns are represented on Figure 8 for three packing fractions in 3D space/time, the time axis being rescaled with respect to the relaxation time $\tau_{1/2} \simeq \tau_\alpha$, the time needed for observing half of the particles having jumped. We draw all cage jumps (black dots) and link the ones separated by a lag time less than τ_{corr}. This defines a network whose vertices are the cage jumps and whose edges are the orientated links towards facilitated jumps. For the loosest packing fraction, all jumps are connected by a facilitation link and form a highly interconnected monolith: $\tau_f \gg \tau_{1/2}$ and dynamical facilitation is conserved on timescales relevant for structural relaxation. When raising ϕ, an increasing number of adjacent clusters become separated by more than a few τ_{corr} within a time interval equal to the relaxation time. Several independent avalanches start and end within a time interval of the order of the relaxation timescale and dynamical facilitation is clearly not conserved anymore. The above observations suggest that at even higher density τ_f would become of the order of τ_{corr}: each avalanche would reduce to a single cluster and dynamical facilitation would disappear completely.

At this point, we see that the spatio-temporal organisation of the cage jumps stands at the root of the dynamical heterogeneities and their evolution when approaching the glass transition. The very strong similarities reported among the granular systems and the repulsive liquid provide a good physical ground for the existence of a universal mechanism ruling the glass transition in these very different systems. Dynamical facilitation should be a key ingredient in these models. If the observations just reported in the case of the FBE also hold in the case of the liquids – which remains to be checked – then coarse grained models should not impose the conservation of facilitation.

3.3. Relation to the underlying structure

A natural question [27] related to these findings is what, if any, structural features are correlated with the heterogeneity noted in the real space dynamics. Important progress in this direction has been obtained [57], through the introduction of the quantitative notion of "propensity", and then later in [59, 12, 61, 62], where it has been shown that irreversible motion is correlated with the spatial characteristics of soft modes.

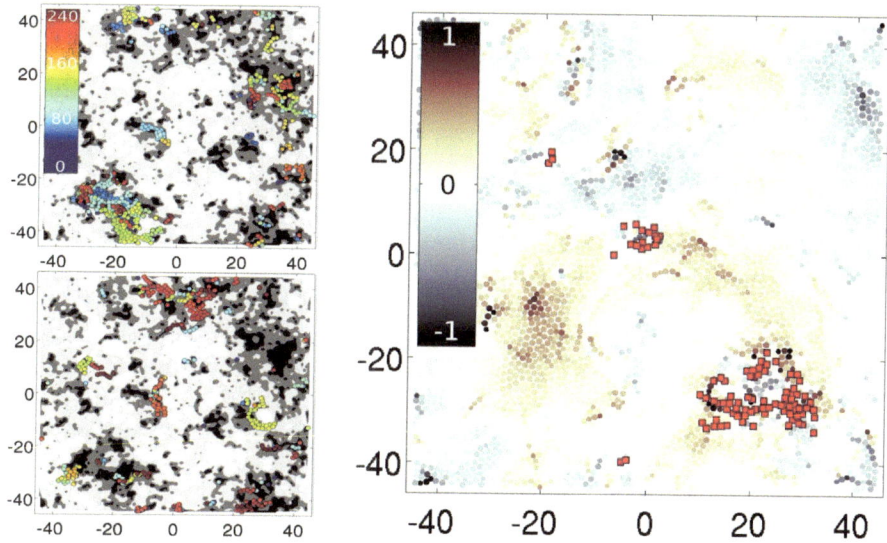

FIGURE 9. Left: Cage jumps occurring between t (blue) and $t + \tau_{corr}$ (red) for two different isoconfigurational trajectories, on top of a DW factor map computed at time t (in grey). Right: Cage jumps occurring in $\tau = 17$ on top of a map of the relative difference $(DW(t + \tau) - DW(t))/\langle DW \rangle$ (from [15]).

Characterizing the structure of amorphous media is a difficult task. Identifying the kind of order, if any, hidden inside the amorphous structure has not been achieved yet. For 2D systems, several authors inspired by solid state approaches propose to define defects in the neighborhood structure. Aharanov *et al.* [2] for instance could provide evidences of the existence of two liquid-like and glass-like defects in a bidisperse soft repulsive disks system. In this case, the glass transition would coincide with the density of liquid-like defects going to zero. A variation of the structural role of defects has been proposed by Tanaka and coworkers [52, 54], who looked at the orientational order and observed some correlation with the dynamical heterogeneities. However such a correlation was never observed, neither in the CSE, nor in the FBE and one may suspect that these approaches based on the identification of defects is not as robust as one would expect for a minimal mechanism.

An alternative way of characterizing the structure is to evaluate the dynamical properties averaged over many initial configurations, identical in structure and with no dynamical information: the so-called iso-configurational ensemble[59, 58]. In practice one selects an initial equilibrium configuration, replaces the momentum of the particles by random variables and runs the dynamics. Then computing the Debye-Waller (DW) factor for particle i $DW_i = \langle [\vec{r}_i(t) - \langle \vec{r}_i \rangle_{\delta t}]^2 \rangle_{\delta t, C}$, where the average is performed over the isoconfiguration ensemble and over a short time interval δt, one obtains a measure of the local ability to move, as encoded in the structure only. DW_i is a good estimate of the local softness.

Starting from the same equilibrated configuration, one can compute the long time dynamics of several iso-configurational trajectories, identify the cage jumps and compare their location to the map of the local DW. Comparing the two panels on the left of Figure 9, one sees that the cage jumps are always located on top of high DW areas, that is at the softest place of the system. Also one observes that for each run they cover only a part of the high DW area, and that they take place at different times and places for each trajectory underlining the stochastic character of these events. So the following picture emerges. There are indeed spatial heterogeneities in the structure: the softness is not evenly distributed and cage jumps occur preferentially in the softest place. However this is not a deterministic process. This is a qualitative but nonetheless clear evidence that structural properties are good predictors of dynamics on large enough time and length scale. However one must realize that when cage jumps occur, they also contribute to the renewal of the DW map. And they do not simply relax the softer regions to harder ones. Otherwise the system would harden and the dynamics would not be stationary. Hence it must be quite a non-local and complicated process. The right of Figure 9 indeed shows that the relative variation of DW on a timescale comparable to the duration of the clusters of cage jumps, spread over large regions, while the cage jumps are grouped in more localized clusters as already discussed above. What is mediating the non-local interaction between cage jumps and DWs is an intriguing question. One possibility is that a slowly varying spatial field, like the thermal strain discussed in [60], plays an important role by providing long ranged dynamical interactions.

4. Summary and open questions

Despite the very deep differences between thermal supercooled liquids at equilibrium and athermal granular media in out of equilibrium steady states, we have seen in this note that the jamming of granular media shares strong similarities – exceedingly slow relaxation and aging – with the glassy dynamics of super-cooled liquids.

These similarities extend and find their origin at the microscopic scale. The trajectories are composed of a short time vibrational part, which does not contribute to the structural relaxation, and quasi-instantaneous events, the cage jumps. For both the two granular media experiment CSE and FBE and the repulsive liquid simulations the dynamics occurs via a two-time scale process that gives rise to dynamical heterogeneities and induces macroscopic relaxation. At short times, the particles cooperatively jump within clusters whose sizes are widely distributed. These clustered jumps trigger other ones nearby, in a facilitated process, leading to large scale avalanches. This organization of the dynamics is strikingly similar despite the differences between the systems considered here. This is related to the separation of the dynamics into well-identified vibrations and cage jumps. whether such processes would also be identified in simulations of other super-cooled liquids like Lennard Jones liquids, or Silica as well as in colloidal suspensions is an obvious question and a quite easy one to answer in the light of the existing data.

In the case of the granular FBE, the dynamical facilitation becomes less conserved and plays a smaller role when increasing the packing fraction towards the glass transition; whether the same scenario holds for thermal liquids is a key issue. Studying the evolution of dynamical properties with decreasing temperature following the same analysis would allow for direct tests of prominent theories of the glass transition. For example, in the picture based on kinetically constrained models of glasses [30] facilitation should become more relevant and conserved upon lowering the temperature. In the Random First-Order Transition Theory [51], the dynamics should be correlated with soft regions for moderately supercooled liquids but, closer to the glass transition, the relaxation should be dominated by other processes.

Finally in the case of the liquid, we have seen that dynamical facilitation is clearly coupled to the structure: mobility preferentially follows the soft regions and has a non-local influence on the evolution of the topography of hard and soft areas. The resulting picture of facilitation is quite different from the one based on the propagation of a conserved mobility field. What are the physical mechanisms at stake in this strong coupling? Are they as general as the cage jumps or does dissipation play a specific role in the case of granular media. These are certainly important questions for the near future.

Acknowledgment

The content of this chapter owes a lot to my collaborators, without whom none of the present ideas and results would have been developed: G. Marty, F. Lechenault, R. Candelier, G. Biroli and J.-P. Bouchaud. Several papers written with them have inspired the present notes. I want to thank A. Abate, D. Durian, A. Widmer-Cooper and P. Harrowell for having shared data with us. I also acknowledge all the members of the "Glassy Work group" in Saclay, for many inspiring discussions. Finally let me express my gratitude to B. Duplantier for giving me the opportunity to include these notes in the present collection.

References

[1] A.R. Abate and D.J. Durian, *Approach to jamming in an air-fluidized granular bed,* Phys. Rev. E **74** (3) (2006), 031308.

[2] E. Aharonov, E. Bouchbinder, HGE Hentschel, V. Ilyin, N. Makedonska, I. Procaccia, and N. Schupper, *Glass Transition and Plasticity,* EPL (*Europhysics Letters*) **77** (2007), 56002.

[3] C.A. Angell, *The Old Problems of Glass and the Glass Transition, and the Many New Twists.* Proceedings of the National Academy of Sciences **92** (15) (1995), 6675–6682.

[4] H.A. Barnes, *The Yield Stress – a Review or 'panta rei' – Everything Flows?* Journal of Non-Newtonian Fluid Mechanics **81** (1999), 133–178.

[5] L. Berthier, G. Biroli, J.P. Bouchaud, L. Cipelletti, D. El Masri, D. L'Hote, F. Ladieu, and M. Pierno, *Direct experimental evidence of a growing length scale accompanying the glass transition*, Science (Washington, D. C.) **310** (5755) (2005), 1797–1800.

[6] L. Berthier and T. A. Witten, *Compressing nearly hard sphere fluids increases glass fragility*, 2009.

[7] L. Berthier and T.A. Witten, *Glass transition of dense fluids of hard and compressible spheres*, Physical Review E **80** (2) (2009), 21502.

[8] S.M. Bhattacharyya, B. Bagchi, and P.G. Wolynes, *Facilitation, complexity growth, mode coupling, and activated dynamics in supercooled liquids*, Proceedings of the National Academy of Sciences **105** (42) (2008),16077.

[9] G. Biroli, J.-P. Bouchaud, A. Cavagna, T.S. Grigera, and P. Verrocchio, *Thermodynamic signature of growing amorphous order in glass-forming liquids*, Nature Physics **4** (2008), 771.

[10] G. Biroli and J.P. Bouchaud, *Diverging length scale and upper critical dimension in the Mode-Coupling Theory of the glass transition*, Europhys. Lett **67** (1) (2004), 21–27.

[11] S. D. Bond, B. J. Leimkuhler, and B. B. Laird, *The nosé-poincaré method for constant temperature molecular dynamics*, Journal of Computational Physics **151** (1999), 114–134.

[12] C. Brito and M. Wyart, *Heterogeneous dynamics, marginal stability and soft modes in hard sphere glasses*, J. Stat. Mech. **8** (2007), L08003.

[13] R. Candelier, O. Dauchot, and G. Biroli, *Building blocks of dynamical heterogeneities in dense granular media*, Physical Review Letters **102** (8) (2009), 088001.

[14] R. Candelier, O. Dauchot, and G. Biroli, *Dynamical facilitation in dense granular media decreases when approaching the glass transition*, Condmat 0912.0472, (2010).

[15] R. Candelier, A. Widmer-Cooper, J. K. Kummerfeld, O. Dauchot, G. Biroli, P. Harrowell, and D. R. Reichman, *Avalanches and dynamical correlations in supercooled liquids*, arXiv.org:0912.0193, (2009).

[16] M.E. Cates, J.P. Wittmer, J.P. Bouchaud, and P. Claudin, *Jamming, Force Chains, and Fragile Matter*, Phys. Rev. Lett. **81** (9) (1998), 1841–1844.

[17] P. Chaudhuri, L. Berthier, and W. Kob, *Universal nature of particle displacements close to glass and jamming transitions*, Physical Review Letters **99** (6) (2007), 60604.

[18] P. Chaudhuri, L. Berthier, and S. Sastry, *Jamming transitions in amorphous packings of frictionless spheres occur over a continuous range of volume fractions*, Arxiv preprint arXiv:0910.0364, (2009).

[19] A.D.L. Cipelletti, *Length scale dependence of dynamical heterogeneity in a colloidal fractal gel*, Europhys. Lett **76** (5) (2006), 972.

[20] O. Dauchot, *Glassy behaviours in a-thermal systems, the case of granular media: A tentative review*, In Michel Pleimling Roland Sanctuary Malte Henkel, editor, *Ageing and the Glass Transition*, Chapter 4,, page 161. Lecture Notes in Physics, Springer Verlag, New-York, Berlin, 2007.

[21] O. Dauchot, G. Marty, and G. Biroli, *Dynamical Heterogeneity Close to the Jamming Transition in a Sheared Granular Material*, Phys. Rev. Lett. **95** (26) (2005), 265701.

[22] P.G. Debenedetti and F.H. Stillinger, *Supercooled liquids and the glass transition*, Nature **410** (6825) (2001), 259–67.

[23] B. Doliwa and A. Heuer, *Cage Effect, Local Anisotropies, and Dynamic Heterogeneities at the Glass Transition: A Computer Study of Hard Spheres*, Phys. Rev. Lett. **80** (22) (1998), 4915–4918.

[24] C. Donati, S. Franz, S.C. Glotzer, and G. Parisi, *Theory of non-linear susceptibility and correlation length in glasses and liquids*, Journal of Non-Crystalline Solids **307** (2002), 215–224.

[25] J.A. Drocco, M.B. Hastings, C.J.O. Reichhardt, and C. Reichhardt, *Multiscaling at Point J: Jamming is a Critical Phenomenon*, Phys. Rev. Lett. **95** (8) (2005), 88001.

[26] M.A. Ediger, Annu. Rev. Phys. Chem. **51** (2000), 99.

[27] M.D. Ediger, C.A. Angell, and S.R. Nagel, *Supercooled liquids and glasses*, J. Phys. Chem. **100** (31) (1996), 13200–13212.

[28] W.G. Ellenbroek, E. Somfai, M. van Hecke, and W. van Saarloos, *Critical Scaling in Linear Response of Frictionless Granular Packings near Jamming*, Phys. Rev. Lett. **97** (25) (2006), 258001.

[29] Glenn H. Fredrickson and Hans C. Andersen, *Kinetic ising model of th glass transition*, Physical Review Letters **53** (1984), 1244.

[30] Juan P. Garrahan and David Chandler, *Geometrical explanation and scaling of dynamical heterogeneities in glass forming systems*, PRL **89** (2002), 035704.

[31] S.C. Glotzer, V.N. Novikov, and T. Schröder, *Time-dependent, four-point density correlation function description of dynamical heterogeneity and decoupling in supercooled liquids*, The Journal of Chemical Physics **112** (2) (2000), 509.

[32] G.Williams and D.C.Watts, Trans. Faraday Soc. **66** (1970), 80.

[33] MM Hurley and P. Harrowell, *Kinetic structure of a two-dimensional liquid*, Physical Review E **52** (2) (1995), 1694–1698.

[34] A. Kabla and G. Debrégeas, *Contact Dynamics in a Gently Vibrated Granular Pile*, Phys. Rev. Lett. **92** (3) (2004), 35501.

[35] A.S. Keys, A.R. Abate, S.C. Glotzer, and D.J. Durian, *Measurement of growing dynamical length scales and prediction of the jamming transition in a granular material*, Nature Physics **3** (4) (2007), 260–264.

[36] J.B. Knight, C.G. Fandrich, C.N. Lau, H.M. Jaeger, and S.R. Nagel, *Density relaxation in a vibrated granular material*, Phys. Rev. E **51** (5) (1995), 3957–3963.

[37] W. Kob, C. Donati, S.J. Plimpton, P.H. Poole, and S.C. Glotzer, *Dynamical heterogeneities in a supercooled Lennard-Jones liquid*, Physical Review Letters **79** (15) (1997), 2827–2830.

[38] R. Kohlrausch, Pogg. Ann. Phys. Chem. **91** (1854), 179.

[39] F. Lechenault, O. Dauchot, G. Biroli, and JP Bouchaud, *Critical scaling and heterogeneous superdiffusion across the Jamming transition*, Europhysics Letters **83** (2008), 46003.

[40] A.J. Liu and S.R. Nagel, *Jamming is not just cool anymore*, Nature **396** (1998), 21–22.

[41] G. Marty and O. Dauchot, *Subdiffusion and Cage Effect in a Sheared Granular Material*, Phys. Rev. Lett. **94** (1) (2005), 15701.

[42] M. Nicolas, P. Duru, and O. Pouliquen, *Compaction of a granular material under cyclic shear*, The European Physical Journal E-Soft Matter **3** (4) (2000), 309–314.

[43] E.R. Nowak, J.B. Knight, E. Ben-Naim, H.M. Jaeger, and S.R. Nagel, *Density fluctuations in vibrated granular materials*, Phys. Rev. E **57** (2) (1998), 1971.

[44] C.S. O'Hern, S.A. Langer, A.J. Liu, and S.R. Nagel, *Random Packings of Frictionless Particles*, Phys. Rev. Lett. **88** (7) (2002), 75507.

[45] C.S. O'Hern, L.E. Silbert, A.J. Liu, and S.R. Nagel, *Jamming at zero temperature and zero applied stress: The epitome of disorder*, Phys. Rev. E **68** (1) (2003), 11306.

[46] P. Olsson and S. Teitel, *Critical Scaling of Shear Viscosity at the Jamming Transition*, Phys. Rev. Lett. **99**, 178001 (2007).

[47] G. Parisi and F. Zamponi, *Mean field theory of the glass transition and jamming of hard spheres*, Reviews of Modern Physics, **82** (1) (2010), 789–845.

[48] P. Philippe and D. Bideau, *Compaction dynamics of a granular medium under vertical-tapping*, Europhysics Letters **60** (5) (2002), 677–683.

[49] P. Philippe and D. Bideau, *Granular medium undervertical tapping: Change of compaction and convection dynamics around the liftoff threshold*, Phys. Rev. Lett. **91** (2003), 104302.

[50] M. Schröter, D.I. Goldman, and H.L. Swinney, *Stationnary state volume fluctuations in a granular medium*, Phys. Rev. E **71** (2005), (030301(R)).

[51] J.D. Stevenson and P.G. Wolynes, *Thermodynamic- Kinetic Correlations in Supercooled Liquids: A Critical Survey of Experimental Data and Predictions of the Random First-Order Transition Theory of Glasses*, J. Phys. Chem. B **109** (31) (2005), 15093–15097.

[52] H. Tanaka, *Two-order-parameter description of liquids. I. A general model of glass transition covering its strong to fragile limit*, The Journal of Chemical Physics **111** (1999), 3163.

[53] M. Van Hecke, *Jamming of Soft Particles: Geometry, Mechanics, Scaling and Isostaticity*, J. Phys.: Condens. Matter, 22:033101, 2010.

[54] K. Watanabe and H. Tanaka. Direct observation of medium-range crystalline order in granular liquids near the glass transition. *Physical review letters*, 100(15):158002, 2008.

[55] E.R. Weeks, J.C. Crocker, A.C. Levitt, A. Schofield, and D.A. Weitz. Three-Dimensional Direct Imaging of Structural Relaxation Near the Colloidal Glass Transition. *Science*, 287(5453):627–631, 2000.

[56] E.R. Weeks and D.A. Weitz. Properties of Cage Rearrangements Observed near the Colloidal Glass Transition. Phys. Rev. Lett. **89** (9) (2002), 95704.

[57] A. Widmer-Cooper and P. Harrowell, *On the relationship between structure and dynamics in a supercooled liquid*, Journal of Physics: Condensed Matter **17** (2005), S4025–S4034.

[58] A. Widmer-Cooper and P. Harrowell, *On the study of collective dynamics in supercooled liquids through the statistics of the isoconfigurational ensemble*, The Journal of chemical physics **126** (2007), 154503.

[59] Adam Widmer-Cooper, Heidi Perry, Peter Harrowell, and David R. Reichman, *Irreversible reorganization in a supercooled liquid originates from localized soft modes,* Nature Physics **4** (2008), 711.

[60] Asaph Widmer-Cooper and Peter Harrowell, *The central role of thermal collective strain in the relaxation of structure in a supercooled liquid,* (2009).

[61] R. Yamamoto and A. Onuki, J. of the Physics Society of Japan **66** (1997), 2545.

[62] R. Yamamoto and A. Onuki, *Dynamics of highly supercooled liquids: Heterogeneity, rheology, and diffusion,* Physical Review E **58** (3) (1998), 3515–3529.

Olivier Dauchot
SPEC, CEA-Saclay
URA 2464 CNRS
L'Orme des Merisiers
F-91191 Gif-sur-Yvette, France
e-mail: `olivier.dauchot@cea.fr`

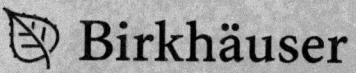
Progress in Mathematics

Poincaré Seminar

The Poincaré Seminar is held at the Institute Henri Poincaré in Paris. The goal of this seminar is to provide up-to-date information about general topics of great interest in physics. Both theoretical and experimental results are covered, with some historical background. Particular care is devoted to the pedagogical nature of presentation.

■ **PMP 60: Duplantier, B. / Rivasseau, V. (eds.)**
Biological Physics. Poincaré Seminar 2009 (2010).
ISBN 978-3-0346-0427-7

This new volume in the Poincaré Seminar Series, describing recent developments at the interface between physics and biology, is directed towards a broad audience of physicists, biologists, and mathematicians. Both the theoretical and experimental aspects are covered, and particular care is devoted to the pedagogical nature of the presentations. Competent contributors describe the theoretical advances made in the study of "active gels", with applications to liquid crystals and cell motility; report on recent advances made with biomimetic model systems in the understanding of cytokinesis; present several molecular models for motor proteins; describe the latest advances made in the real-time single molecule study of the enzymes involved in DNA replication; address the problem of understanding, from a physics perspective, the driving forces behind the biological evolution of multicellularity; and more.

■ **PMP 55: Duplantier, B. / Raimond, J.-M. / Rivasseau, V. (eds.)** The Spin. Poincaré Seminar 2007 (2009).
ISBN 978-3-7643-8798-3

The Poincaré Seminar "The Spin" focussed on how this once mysterious quantum reality called spin has become ubiquitous in modern physics from the most theoretical aspects down to the most practical applications.
The first and more theoretical part of the book starts with a detailed presentation of the notion

of spin by leading world expert Jürg Fröhlich. He reviews its historical development in quantum mechanics and its increasing relevance to quantum field theory and condensed matter. The next two chapters, by Nobel laureate Franck Wilczek and Stephane Ouvry, discuss the exotic anyon particles. The second part of the book, with contributions by Gerald Gabrielse, Nobel laureate Albert Fert and his collaborators Pierre Sénéor, Vincent Cros and Frédéric Petroff as well as Pierre-Jean Nacher, aims at the presentation of the most advanced current experiments and applications of the notion of spin.

■ **PMP 53: Damour, T. / Duplantier, B. / Rivasseau, V. (eds.)**
Quantum Spaces. Poincaré Seminar 2007 (2007). ISBN 978-3-7643-8521-7

■ **PMP 52: Damour, T. / Duplantier, B. / Rivasseau, V. (eds.)** Gravitation and Experiment. Poincaré Seminar 2006 (2007).
ISBN 978-3-7643-8523-1

■ **PMP 48: Duplantier, B. / Raimond, J.-M. / Rivasseau, V. (eds.)**
Quantum Decoherence. Poincaré Seminar 2005 (2006). ISBN 978-3-7643-7807-3

■ **PMP 47: Damour, T. / Darrigol, O. / Duplantier, B. / Rivasseau, V. (eds.)**
Einstein, 1905–2005. Poincaré Seminar 2005 (2006). ISBN 978-3-7643-7435-8

■ **PMP 45: Douçot, B. / Duplantier, B. / Pasquier, V. / Rivasseau, V. (eds.)**
The Quantum Hall Effect. Poincaré Seminar 2004 (2005). ISBN 978-3-7643-7300-9

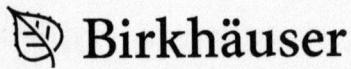 **Birkhäuser** | **birkhauser-science.com**

Progress in Mathematical Physics

Progress in Mathematical Physics is a book series encompassing all areas of theoretical and mathematical physics. It is intended for mathematicians, physicists, and other scientists, as well as graduate students in the above related areas. This distinguished collection of books includes authored monographs and textbooks, the latter primarily at the senior undergraduate and graduate levels. Edited collections of articles on important research developments or expositions of particular subject areas may also be included.

Edited by
Anne Boutet de Monvel (Université Paris VII Denis Diderot, France)
Gerald Kaiser (Center for Signals and Waves, Austin, TX, USA)

■ **PMP 60: Duplantier, B. / Rivasseau, V. (eds.)**
Biological Physics. Poincaré Seminar 2009 (2010).
ISBN 978-3-0346-0427-7

■ **PMP 59: Torres de Castillo, G.F.**
Spinors in Four-Dimensional Spaces (2010).
ISBN 978-0-8176-4983-8

■ **PMP 58: Girbau, J. / Bruna, L.**
Stability by Linearization of Einstein's Field
Equation (2010).
ISBN 978-3-0346-0303-4

■ **PMP 57: Volkenstein, M.V.**
Entropy and Information (2009).
ISBN 978-3-0346-0077-4

■ **PMP 56: Sharan, P.**
Spacetime, Geometry and Gravitation (2009).
ISBN 978-3-7643-9970-2

■ **PMP 55: Duplantier, B. / Raimond, J.-M. /
Rivasseau, V. (eds.)** The Spin. Poincaré Seminar
2007 (2009).
ISBN 978-3-7643-8798-3

■ **PMP 54: de Oliveira, C.R.**
Intermediate Spectral Theory and Quantum
Dynamics (2009). ISBN 978-3-7643-8794-5

■ **PMP 53: Damour, T. / Duplantier, B. /
Rivasseau, V. (eds.)**
Quantum Spaces. Poincaré Seminar 2007 (2007).
ISBN 978-3-7643-8521-7

■ **PMP 52: Damour, T. / Duplantier, B. /
Rivasseau, V. (eds.)** Gravitation and Experiment.
Poincaré Seminar 2006 (2007).
ISBN 978-3-7643-8523-1

■ **PMP 51: Mathúna, D.Ó.**
Integrable Systems in Celestial Mechanics (2008).
ISBN 978-0-8176-4096-5

■ **PMP 50: Anglès, P.**
Conformal Groups in Geometry and Spin
Structures (2007). ISBN 978-0-8176-3512-1

■ **PMP 49: Palmer, J.**
Planar Ising Correlations (2007).
ISBN 978-0-8176-4248-8

■ **PMP 48: Duplantier, B. / Raimond, J.-M. /
Rivasseau, V. (eds.)**
Quantum Decoherence. Poincaré Seminar 2005
(2006). ISBN 978-3-7643-7807-3

■ **PMP 47: Damour, T. / Darrigol, O. / Duplantier,
B. / Rivasseau, V. (eds.)**
Einstein, 1905–2005. Poincaré Seminar 2005
(2006). ISBN 978-3-7643-7435-8

■ **PMP 46: Marchenko, V.A. / Khruslov, E.Y.**
Homogenization of Partial Differential Equations
(2006). ISBN 978-0-8176-4351-5

■ **PMP 45: Douçot, B. / Duplantier, B. / Pasquier,
V. / Rivasseau, V. (eds.)**
The Quantum Hall Effect. Poincaré Seminar 2004
(2005). ISBN 978-3-7643-7300-9

■ **PMP 44: Demuth, M. / Krishna, M.**
Determining Spectra in Quantum Theory (2005)
ISBN 978-0-8176-4366-9

■ **PMP 43: Bermúdez de Castro, A.**
Continuum Thermomechanics (2005).
ISBN 978-3-7643-7265-1